NANOTECHNOLOGY SCIENCE AND TECHNOLOGY

ACHIEVING COMPLETE BAND GAPS USING LOW REFRACTIVE INDEX MATERIAL

NANOTECHNOLOGY SCIENCE AND TECHNOLOGY

Additional books in this series can be found on Nova's website under the Series tab.

Additional E-books in this series can be found on Nova's website under the E-books tab.

NANOTECHNOLOGY SCIENCE AND TECHNOLOGY

ACHIEVING COMPLETE BAND GAPS USING LOW REFRACTIVE INDEX MATERIAL

DAHE LIU
TIANRUI ZHAI
AND
ZHAONA WANG

Novinka
Nova Science Publishers, Inc.
New York

Copyright © 2010 by Nova Science Publishers, Inc.

All rights reserved. No part of this book may be reproduced, stored in a retrieval system or transmitted in any form or by any means: electronic, electrostatic, magnetic, tape, mechanical photocopying, recording or otherwise without the written permission of the Publisher.

For permission to use material from this book please contact us:
Telephone 631-231-7269; Fax 631-231-8175
Web Site: http://www.novapublishers.com

NOTICE TO THE READER
The Publisher has taken reasonable care in the preparation of this book, but makes no expressed or implied warranty of any kind and assumes no responsibility for any errors or omissions. No liability is assumed for incidental or consequential damages in connection with or arising out of information contained in this book. The Publisher shall not be liable for any special, consequential, or exemplary damages resulting, in whole or in part, from the readers' use of, or reliance upon, this material.
Independent verification should be sought for any data, advice or recommendations contained in this book. In addition, no responsibility is assumed by the publisher for any injury and/or damage to persons or property arising from any methods, products, instructions, ideas or otherwise contained in this publication.
This publication is designed to provide accurate and authoritative information with regard to the subject matter covered herein. It is sold with the clear understanding that the Publisher is not engaged in rendering legal or any other professional services. If legal or any other expert assistance is required, the services of a competent person should be sought. FROM A DECLARATION OF PARTICIPANTS JOINTLY ADOPTED BY A COMMITTEE OF THE AMERICAN BAR ASSOCIATION AND A COMMITTEE OF PUBLISHERS.

LIBRARY OF CONGRESS CATALOGING-IN-PUBLICATION DATA
Available upon Request
ISBN: 978-1-61728-685-8

Published by Nova Science Publishers, Inc. ✚ *New York*

CONTENTS

Preface		vii
Chapter 1	Introduction	1
Chapter 2	Complex Diamond Structure	3
Chapter 3	Self-Simulating Structure	15
Chapter 4	A D_{nv} Point Group Structure Based on Heterostructure	21
Chapter 5	Theoretical Investigation	31
Chapter 6	Temperature Tunable Random Lasing in Weakly Scattering Structure Formed by Speckle	41
References		49
Index		55

PREFACE

Increased interest has been focused on fabricating three dimensional (3D) photonic crystals (PCs) in order to obtain CBGs. Theoretical analysis showed that although CBGs can be obtained by diamond structure, a strict condition should be satisfied, i.e., the modulation of the refractive index of the material used should be larger than 2.0. Therefore, attention was paid to finding the materials with high refractive index. Some scientists tried to fill the templates with high refractive index materials to increase the modulation of the refractive index, and CBGs were obtained. However, the CBGs achieved in 3D PCs were mostly in microwave or infrared regions. This book presents and discusses various advances in this field of research.

Chapter 1

INTRODUCTION

Photonic crystals (PCs) was introduced by E. Yablonovitch [1] and S. John [2] in 1987. In 1990, Ho *et al.* demonstrated theoretically that a diamond structure possesses complete band gaps (CBGs) [3]. Since then, great interest was focused on fabricating three dimensional (3D) photonic crystals (PCs) in order to obtain CBGs. Several methods were reported, and CBGs were achieved in the range of microwave or submicrowave. Theoretical analysis showed that although CBGs can be obtained by diamond structure, a strict condition should be satisfied, i.e., the modulation of the refractive index of the material used should be larger than 2.0 [4,5]. So, attention was paid to finding the materials with high refractive index. Some scientists tried to fill the templates with high refractive index materials to increase the modulation of the refractive index [6-8], and CBGs were obtained. However, the CBGs achieved in 3D PCs were mostly in microwave or infrared regions [7,9-11]. Holography is a cheap, rapid, convenient, and effective technique for fabricating 3D structures. In 1997, holographic technique was introduced for fabricating the face centered cubic (fcc) structure [12]. Campbell *et al.* actually fabricated the fcc structure with holographic lithography [13]. Several authors reported their works on this topic [4,5,9,14]. Toader *et al.* also showed theoretically a five-beam "umbrella" configuration in the synthesis of a diamond photonic crystal [15]. Because both the value and modulation of the refractive index of the holographic recording materials are commonly low, there would be no CBGs in PCs made directly by holography. For example, the epoxy photoresist generally used for holographic lithography [13,16-18] has n=1.6, which is a little bit too low. They may, however, be used as templates for the production of inverse replica structures by, for example, filling the void with high refractive index and burning out or dissolving the photoresist [13], and a good work was done by Meisel *et al* [19,20]. However, to find materials with large refractive index is not easy, and the special techniques needed are very complicated and expensive. It limits the applications of PCs, especially for industrial productions.

Although some efforts had been made [20], CBGs in the visible range had not yet been achieved by using the materials with low refractive index. Therefore, it is a big challenge to fabricate 3D PCs possessing CBGs in the visible range by using materials with low refractive index, though it is greatly beneficial for future PC industry. As a first step for achieving this, it is important to obtain very wide band gaps. In the previous investigations [21,22], it was evidently shown that the anisotropy of a photonic band gap in a two-dimensional photonic crystal is dependent on the symmetry of the structure, and as the order of the symmetry increases, it becomes easier to obtain a complete band gap. One would naturally ask whether or not such a method is applicable to 3D PCs. In view of this question, we proposed a series holographic method for fabricating some special structures by using materials with low refractive index, and the features of the band gap in such structures were then studied experimentally and theoretically.

Chapter 2

COMPLEX DIAMOND STRUCTURE [23]

1) EXPERIMENTAL METHOD AND THE SAMPLES

It is known that a cell of diamond structure consists of two cells of fcc structures, and the two cells have a distance of one-quarter of the diagonal length of the cell along the diagonal line. According to this, the diamond structure here was implemented by holography through two exposures: an fcc structure was recorded in the first exposure, then, after the recording material was translated one-quarter of the diagonal length along the diagonal line of the fcc structure, a second exposure was made to record another fcc structure. In this way, a PC with diamond structure was obtained.

Figure 1 shows schematically the optical layout in our experiments. Four beams split from a laser beam were converged to a small area. The central beam was set along the normal direction of the surface of the plate, while the other three outer beams were set around the central beam symmetrically with an angle of 38.9° with respect to the central one. The laser used was a diode pumped laser working at 457.9 nm with linewidth of 200 kHz (Melles Griot model 85-BLT-605). The polarization state of each beam was controlled to achieve the best interference result [24,25]. The recording material was mounted on a one-dimensional translation stage driven by a stepping motor with a precision of 0.05 μm/step. The translation stage was mounted on a rotary stage. The holographic recording material used was dichromated gelatin (DCG) with refractive index n=1.52. The maximum value of its refractive index modulation Δn can reach around 0.1, which is a very small value for obtaining wide band gaps. The thickness T of the material was 36 μm. The DCG emulsion was coated on an optical glass plate with flatness of $\lambda/10$ and without any doping.

It is known that there are several directions with high symmetry in the first Brillouin zone of an fcc structure (see Figure 2). In our experiments, two exposures as mentioned above were made firstly in

Γ-L ([111]) direction to get a diamond structure. Then, a second, even a third, diamond structure was implemented by changing the orientation of the recording material by rotating the rotary stage to other symmetric directions, i.e., the direction of Γ-X ([100]) and/or the direction of Γ-K ([110]) in the first Brillouin zone of the first fcc structure. In this way, photonic crystals with one, two, and even three diamond structures were fabricated. It should be pointed out that when the second or the third diamond structure was recorded, the angle between any two beams should be changed to guarantee that all the beams inside the medium satisfy the relation shown in Figure 1 so that the standard diamond structures can be obtained.

Since the hologram made with DCG is a phase hologram, there is only the distribution of the refractive index but no plastic effect inside the hologram, so a scanning electron microscope (SEM) image cannot be obtained. However, microscopic image can be obtained. To verify the structure in the hologram, the same fcc structure was implemented using the same optical layout but with the photoresist of $2\,\mu m$ thickness. Figure 3 shows the structure made with the photoresist at [111] plane: Figure 3 (a) shows the SEM image and Figure 3(b) shows the optical microscopic image taken with a charge coupled device (CCD) camera mounted on a 1280× microscope. The pixel size of the CCD is $10\,\mu m$. Figure 3(c) shows the optical microscopic image of a diamond structure made with DCG of $36\,\mu m$ at [111] plane, which was also recorded by the CCD camera mounted on a 1280× microscope as mentioned above.

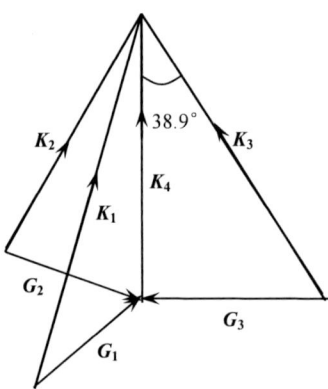

Figure 1. Schematic optical layout for recording an fcc structure.

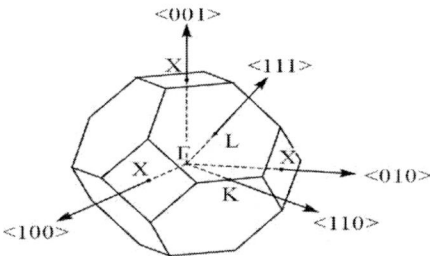

Figure 2. First Brillouin zone of an fcc structure.

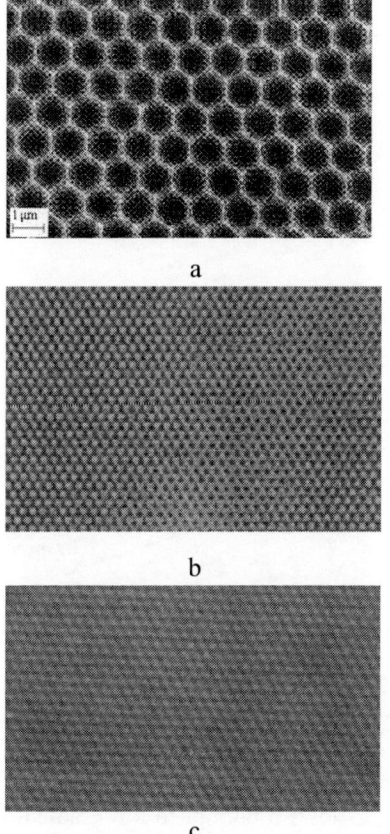

Figure 3. Structure of PCs made in our laboratory. The images (b) and (c) were recorded by a CCD camera mounted on a 1280× microscope. (a) SEM image of an fcc structure made with photoresist of 1 μm thickness. (b) Microscopic image of an fcc structure made with photoresist of 1 μm thickness. (c) Microscopic image of a diamond structure made with DCG of 36 μm.

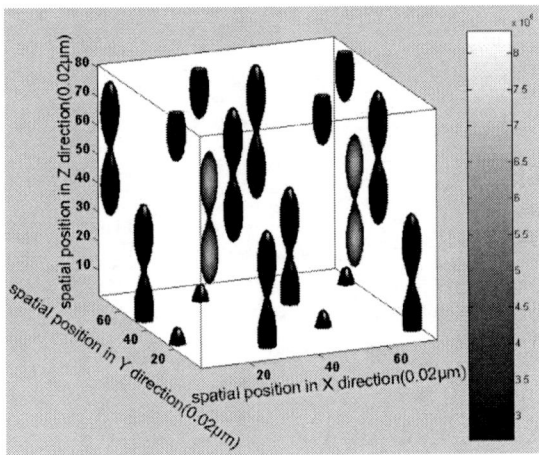

Figure 4. Diamond lattice formed by two fcc lattices in nonlinear exposures. The vertical bar shows a gradient of bottom to top corresponding to the outer to inner region of a cross-sectional cut of a football. The gradient of the outer surface is related to the value of the vertical bar.

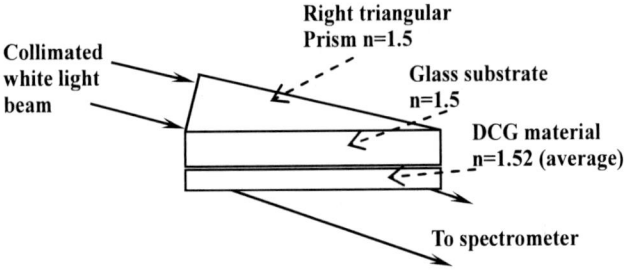

Figure 5. Setup geometry for measuring transmission spectra of photonic crystals.

The theoretical calculated diamond lattice formed by two exposures as mentioned above is shown in Figure 4. It should be pointed out that the shape of the interference results appears like an American football. In this figure, the vertical bar shows a bottom to top gradient corresponding to the outer to inner region of a cross-sectional cut of a football. The gray gradient along the outer surface is related to the value of the vertical bar. Whether a diamond lattice can be obtained depends on how far the two footballs stand apart. If the two footballs are large enough that they overlap, the structure cannot be considered as a diamond lattice. If the two footballs can still be recognized as two, it means that a diamond lattice is formed. In our experiments, the absorption coefficient α of the material was chosen as $\alpha = 1/T$ (Ref. 26) to get optimal interference result. According to previous works, it is well known that for holographic recording materials the relation between the optical density and the exposure is nonlinear. The holographic exposure can be controlled in the region of stronger nonlinear dependence between optical density and exposure as discussed previously [27]. In this way, two fcc lattices can be recognized, so that the structure can be considered as a diamond lattice.

The transmission spectra of PCs with diamond structure made with DCG were measured. In the measurements, a J-Y 1500 monochromator was employed. To minimize the energy loss from reflection at large incident angle, several right triangular prisms were used. The measuring setup geometry is shown in Figure 5. The measured range of the wavelength was 390–800 nm. The [111] plane of the PC measured was set firstly at an arbitrary orientation. The incident angle of the collimated white light beam was changed from 0° to ±90° (the incident angle is in the plane parallel to the surface of the paper). Then, the PC was rotated to other orientations (the orientation angle is in the plane perpendicular to the surface of the paper), and the incident angle of the collimated white light beam was also changed from 0° to ±90°. The orientation angle and the incident angle are the angles φ and θ in a spherical system indeed. In this way, the measurements give actually 3D results.

2) MEASURED SPECTRA AND DISCUSSIONS

The measured transmission spectra of a fabricated triple diamond PC are shown in Figs. 6(a)–6(c), denoting measured results at different orientations of the sample. Each curve in the figures gives the spectrum measured at a certain incident angle and a certain orientation. Different curves give the results at different incident angles and orientations. Thus, whether or not the common gap exists can be determined by the intercept of all the curves. It can be seen that, for the PC with three diamond structures (six fcc structures), the width of the band gaps reached 260 nm, the ratio between the width and the central wavelength of the gaps reached 50 %, and there is a common band gap with a width of about 20 nm at 450–470 nm in the range of 150° of the incident angle. The common gap obtained using DCG with very low refractive index (n=1.52) existed in a wide range, which reached 83 % of the 4π solid angle. Although a complete band gap for all directions was not obtained in our experiments, it is significant to achieve such a wide angle band gap by using a material (DCG) with very low refractive index, because this angular tuning range satisfies most applications in practice, for example, restraining spontaneous radiation with low energy loss in a wide range, wide angle range filter, or reflector with low energy loss.

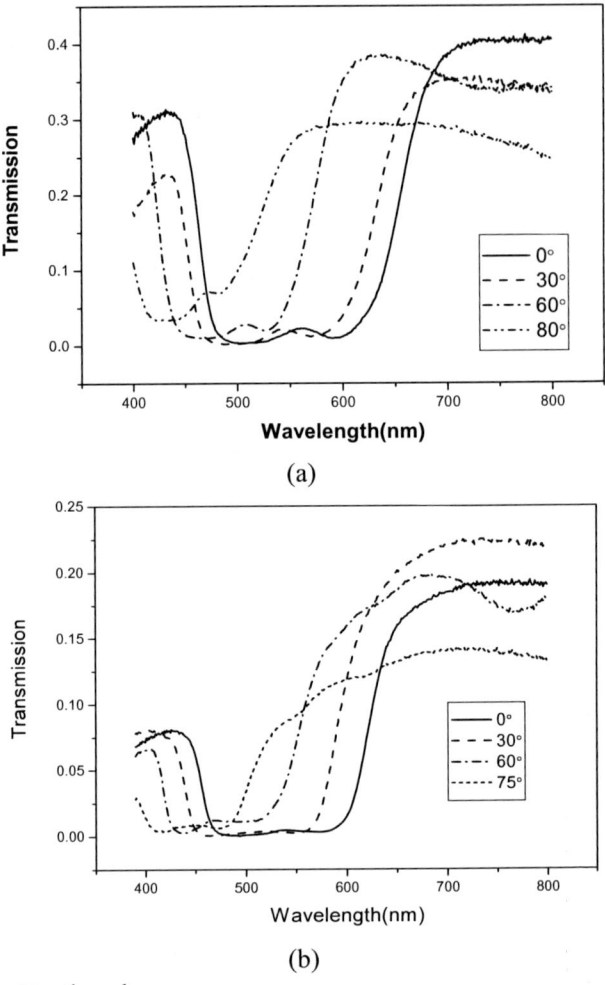

(a)

(b)

Figure Continued

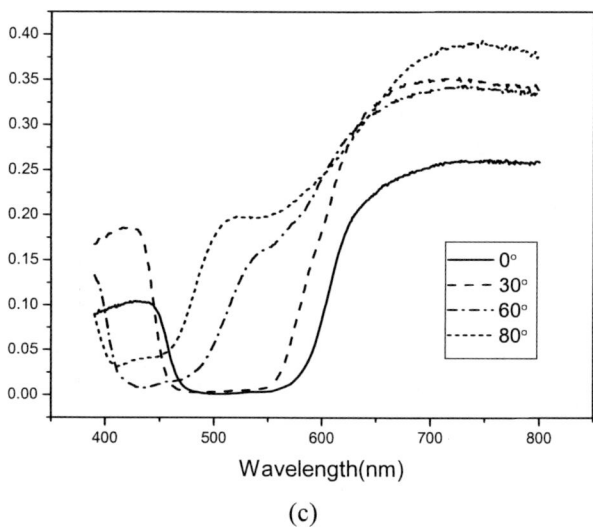

(c)

Figure 6. Measured transmission spectra of PC with three diamond lattices recorded by holography in Γ-L, Γ-X, and Γ-K directions of an fcc lattice respectively. (a), (b), and (c) correspond to the measured results. The angles appearing in (a), (b), and (c) are the incident angles, which are in a plane parallel to the surface of the paper. (a) Orientation angle is 0°. (b) Orientation angle is 30°. (c) Orientation angle is 90°.

This interesting result comes from a complex diamond structure. The photonic crystal with triple-diamond structures is an actual multi structure, but not a stack of several same structures. The [111] direction of the first diamond structure is actually the $\Gamma - X$ direction of the second diamond structure and the $\Gamma - K$ direction of the third diamond structure, respectively. When a beam is incident normally on the PC, the beam is in the [111] direction of the first diamond structure. When the incident angle changes, the beam deviates from the [111] direction of the first diamond structure and tends to approach the [111] direction of the second or third diamond structure. Therefore, though the incident angle changes, the light beam remains always around the [111] direction of the other diamond structures. The narrow region in the $K - \omega$ dispersion relation of a diamond lattice may be expanded by other diamond lattices.

The advantage of a multi diamond structure can be seen more clearly from a comparison with those having one or two diamond structures. Figures 7 and 8 give the measured transmission spectra of PCs with one and two diamond lattices, respectively, at all orientations. The structure with single diamond lattice was fabricated in $\Gamma-L$ [111] direction, while the structure with two diamond lattices was fabricated in $\Gamma-L$ [111] and $\Gamma-X$ [100] directions. Besides, the method used for measuring the two structures with single and double diamond lattices, respectively, was similar to that mentioned above. It can be found that, for the single diamond structure, the space angle range of the common gap is 40°; for the double diamond structures, it becomes 80°. Comparing Figs. 6–8, it is obvious that the width of the band gaps of a PC can be broadened effectively by increasing the number of diamond lattices. This means that the common gap can be enlarged by means of multi structures. The physical origin of such a phenomenon can be understood from the change of structure symmetry. With the increase of the number of diamond lattices, the symmetries around the center of the structure become higher. This made it easier to obtain a broader response in many directions (common band gap in a wide range of angles), which is similar to the case of two dimensional PCs (see Refs. 19 and 20).

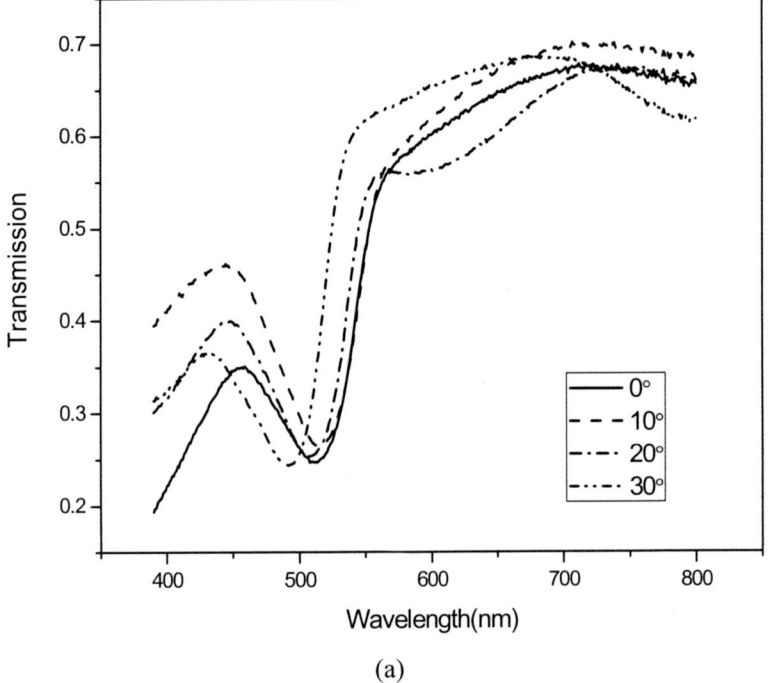

(a)

Figure continued

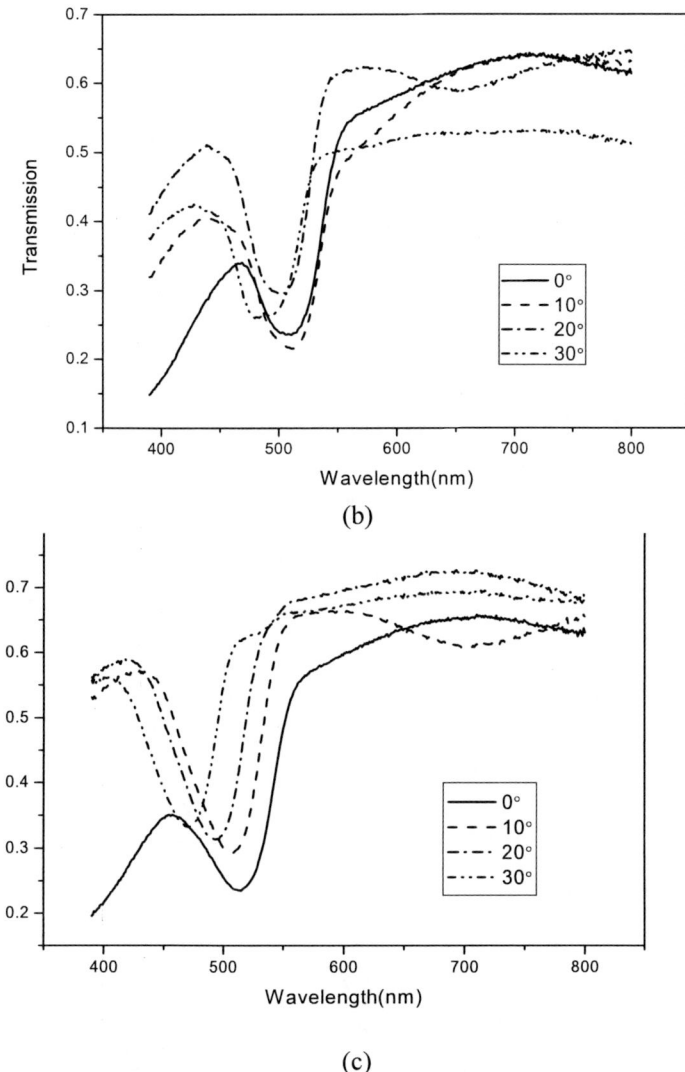

(b)

(c)

Figure 7. Measured transmission spectra of PC with single diamond lattice recorded by holography in $\Gamma-L$ direction of an fcc lattice. (a), (b) and (c) correspond to the measured results. The angles appeared in (a), (b) and (c) are the incident angle which is in a plane parallel to the surface of the paper. (a) Orientation angle is $0°$. (b) Orientation angle is $30°$. (c) Orientation angle is 90^0.

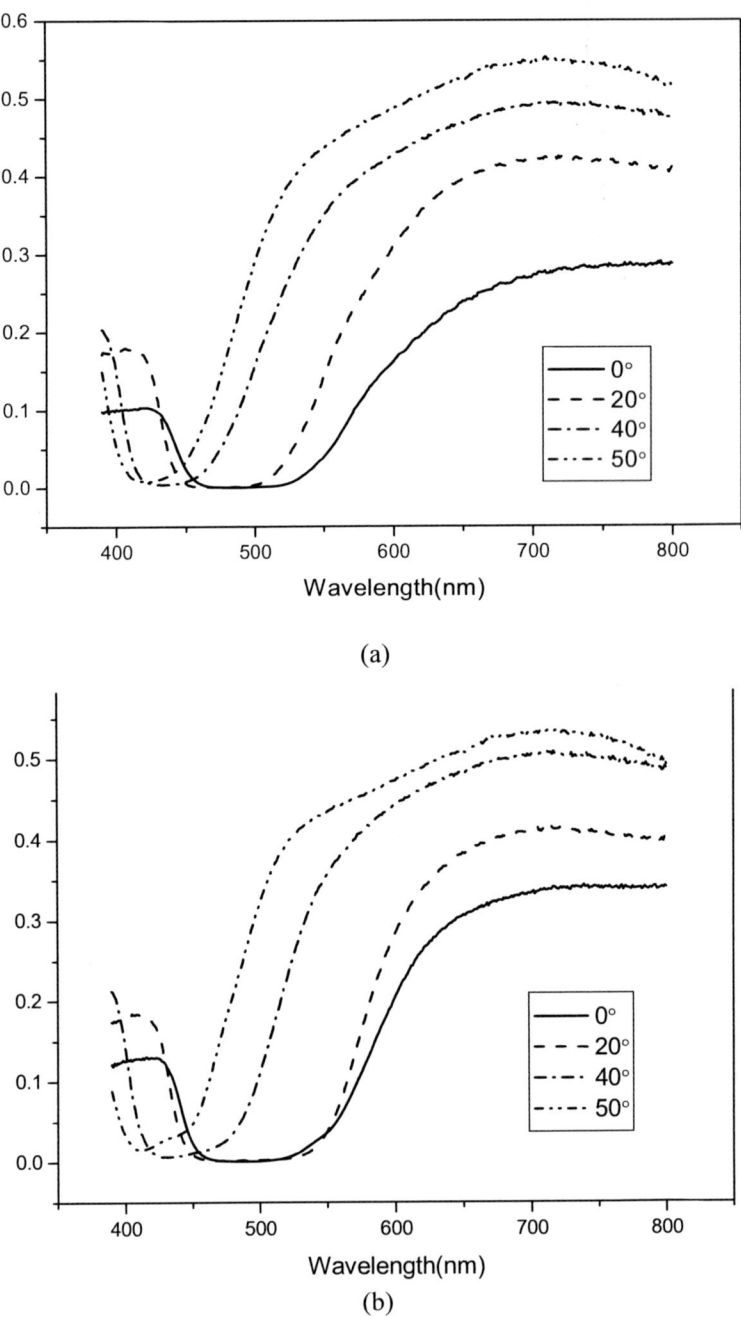

(a)

(b)

Figure Continued

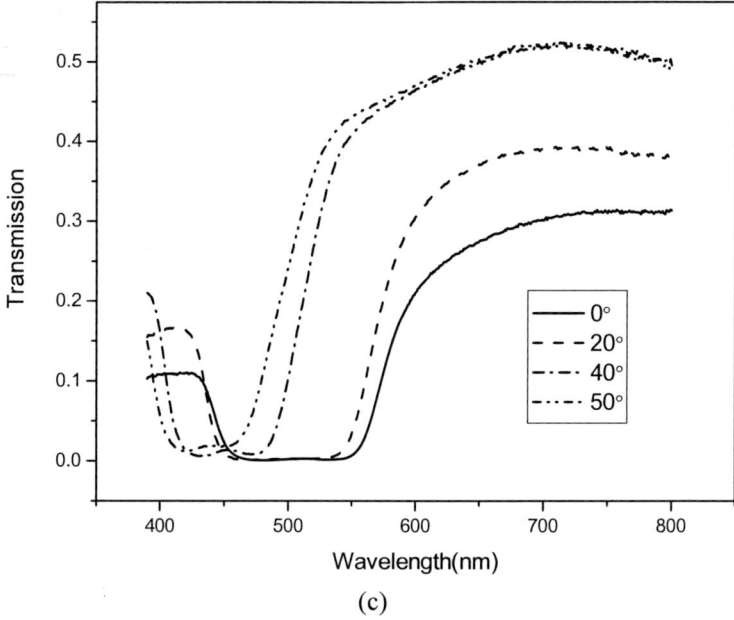

(c)

Figure 8. Measured transmission spectra of PC with 2 diamond lattices recorded by holography in $\Gamma-L$ and $\Gamma-X$ directions of an fcc lattice respectively. (a), (b) and (c) correspond to the measured results. The angles appeared in (a), (b) and (c) are the incident angle which is in a plane parallel to the surface of the paper.

(a) Orientation angle is 0°.

(b) Orientation angle is 30°.

(c) Orientation angle is 90°.

3) CONCLUSION

The width of the band gaps in a PC made by a material with low refractive index can be broadened greatly by multi diamond structures, and a common band gap in the range of 150° of the incident angle can be obtained. This technique will be greatly beneficial in achieving complete band gaps by using materials with low refractive index.

Chapter 3

SELF-SIMULATING STRUCTURE [28,29]

As mentioned above, a complex diamond lattice was fabricated using a materials with low refractive indices. Although a common gap in the visible range was obtained in a wide range that reached 83 % of the 4π solid angle [23], the CBGs was not yet obtained. However, regardless of whether or not they succeed, the above mentioned techniques are complicated and difficult to carry out. Now, a novel technique is developed by which a wide CBG can be achieved easily in the visible range using materials with a low refractive index.

It is obvious that, for a self-simulating spherical system shown in Figure 9, when a light beam passes through the center of the sphere, it will be symmetric in all directions with respect to the center. Therefore, if a PC has this self-simulating spherical structure, it will possess complete band gaps. It should be emphasized that achieving CBGs using this structure does not pose any special requirement on the refractive index of the material used. This type of self-simulating spherical structures can be implemented easily via holography. This includes three steps: 1) preparation of the recording material by coating the emulsion on a soft substrate; 2) fabrication of a 1D PC in rectangular coordinates by using this recording material; 3) to make this PC into a sphere. In this way, a 1D PC in rectangular coordinates was transformed to a spherical 1D PC system.

Figure 10 shows the experimental set-up geometry. Dichromated gelatin (DCG) was still used as recording material. The recording material was prepared by coating a thick layer of 36 mm dichromated gelatin (DCG) onto polyethylene terephthalate (PET). The refractive index of the DCG is 1.52, and the refractive index of the PET is 1.64. Figure 6 shows the optical layout. The laser used was a diode-pumped laser working at 457.9 nm with a line width of 200 KHz (Melles Griot model 85-BLT 605). The monochrometer used was a J–Y 1500. The light beam for measuring was a narrow beam with a diameter of 3 mm. By controlling the wavelength of the laser and the angle between the two beams incident on both sides of the recording material, the period of the PC could be changed for working in the visible, IR, or UV range. Because of the influences of the absorption by the recording material [30], and some factors in processing after exposure of the recording material [31], the structure of the PC in rectangular coordinates is not strictly periodical. Figure 11 shows the structure and distribution of the modulation Δn of the refractive index inside the PC. When the PC in rectangular coordinates is transformed to a sphere, its structure is just the self-simulating spherical system shown in Figure 9.

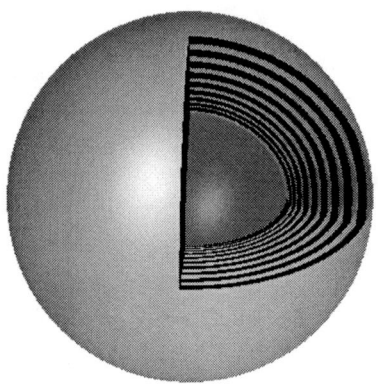

Figure 9. Self-simulating spherical structure.

Several self-simulating spherical structures with different diameters were fabricated. All of them showed identical characteristics under the same experimental conditions. The measured transmission and reflection spectra of a self-simulating spherical structure with the radius of 20 mm were demonstrated as follows. During the measurements, the angles θ and φ of the spherical system were changed from $0°$ through $180°$ and from $0°$ through $360°$, respectively. Figs. 12 and 13 give only the measured results for φ changes from $0°$ through $360°$ while θ was set at a fixed value of $90°$. When the value of θ is changed in the range of $0-180°$, the measured transmission and reflection spectra are quite consistent. For comparison, the theoretical results calculated by transfer matrix method are also shown.

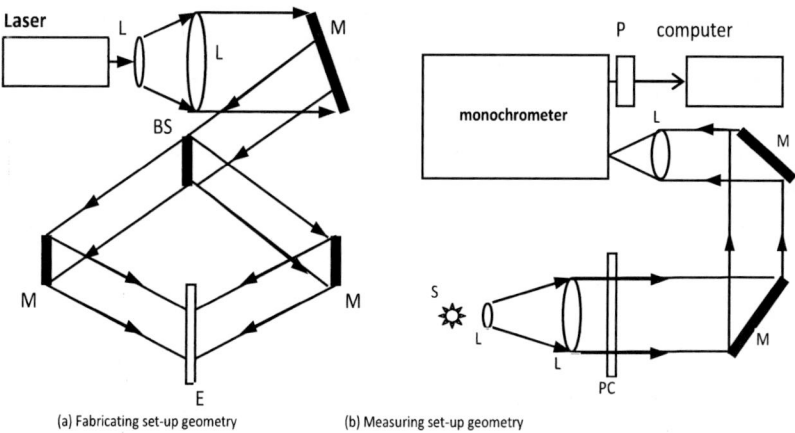

Figure 10. Optical layout for fabricating and measuring PC. L is lens, M is mirror, BS is beam splitter, E is recording material, S is white light source, PC is photonic crystal, P is photomultiplier tube.

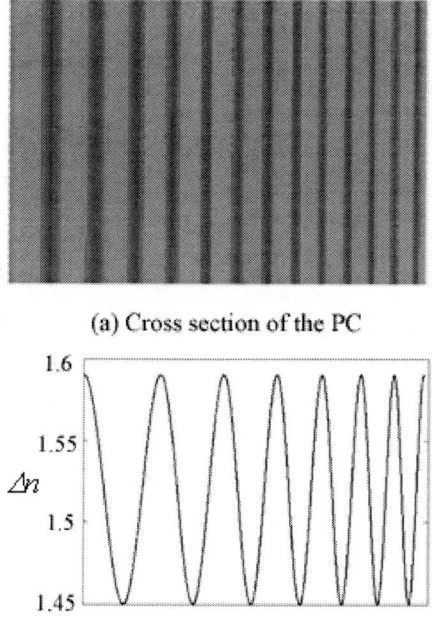

Figure 11. Structure of the fabricated PC in rectangular coordinates.

From Figure 12, it can be seen clearly that, for different values of w in the range of 0–360°, the measured transmission spectra almost completely overlapped with little frequency shift. The width of the CBG reaches 120 nm although the refractive index of DCG is pretty low. The reason is that a structure with non-strict period can broaden the band gaps effectively [32]. Figure 13 shows that the reflection spectra are just complementary to the transmission spectra.

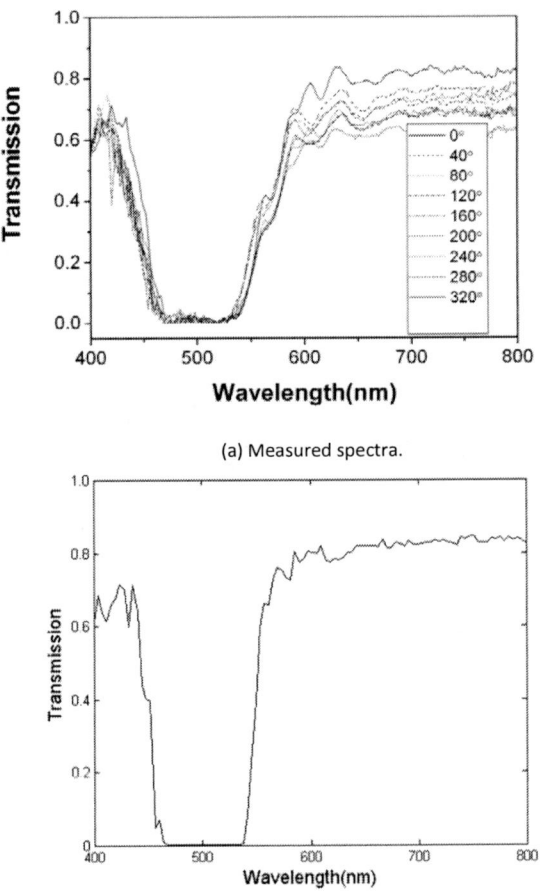

(a) Measured spectra.

(b) Calculated spectra.

Figure 12. Measured and calculated transmission spectra of a self-simulating spherical structure for different value of φ at $\theta = 90°$.

In practical experiments and applications, frequently, the incident beam may not pass through the center of the self-simulating spherical structure strictly. It may affect the characteristics of the band gaps. Figure 13 show the measured transmission spectra when the incident beam deviates from the center of the self-simulating spherical structure. It can be seen that, even though the deviation is as large as 15 mm from the center of the sphere with a radius of 20 mm, there is still a complete band gap with a width of 30 nm. This deviation corresponds to the condition that a light beam incidents on the structure with an incident angle of 49°.

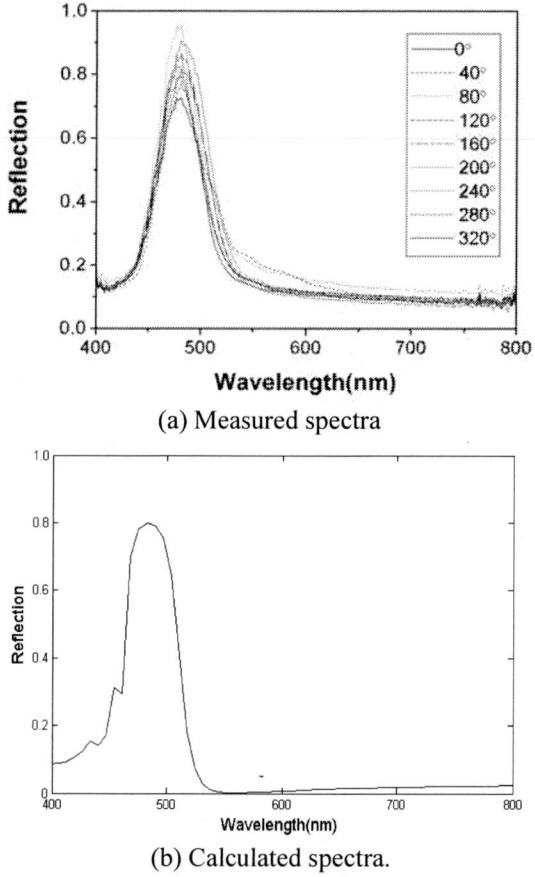

(a) Measured spectra

(b) Calculated spectra.

Figure 13. Measured and calculated reflection spectra of a self-simulating spherical structure for different value of φ at $\theta = 90°$

Self-simulating spherical structures with radii of 5 mm, 7.5 mm, 10 mm, 20 mm, and 25 mm were fabricated and measured in our laboratory, and similar results to those shown in Figs. 12 and 13 were obtained. The measured transmission and reflection spectra revealed that a true complete band gap actually exists for this kind of structure. The fact that the light was forbidden from passing through the structure is really caused by the band gaps, rather than induced by other physical mechanisms, for example, total inner reflection and absorption. An actual fabricated Self-simulating spherical structure is shown in Figure 14.

In summary, a self-simulating spherical structure possesses complete band gaps. This kind of structure can be implemented easily in terms of holography. Achieving complete band gaps with this structure can use common holographic recording materials with low refractive indices rather than high refractive index materials. Even though the working condition has a relative large deviation from ideal conditions, there is still a complete band gap.

Chapter 4

A D_{NV} POINT GROUP STRUCTURE BASED ON HETEROSTRUCTURE [33]

Although the self-simulating spherical structures can obtain CBG, the incident beam must pass through the center of the sphere. however, when light incidents on the surface of the sphere with a incident angle larger than 49°, the CBG will disappear. This disadvantage limits the application of the structure, such as the use of wide incident beam in practical applications. So, improvement of the self-simulating spherical structures is needed.

It was demonstrated that PC heterostructure can improve effectively the characteristics of PCs, and can provide a necessary variation in material properties to turn a photonic crystal raw material into a functional device [34] by means of self-organizing [35-37], lithography with electron beam [38,39] or autocloning [40]. PC heterostructures, like their semiconductor counterparts which are made by combining at least two materials that have distinct band structures, can be fabricated by changing the lattice constant, hole size, or even lattice geometry in the crystal. This can be done either abruptly or gradually [34].

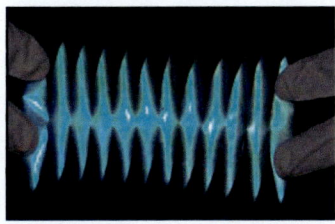 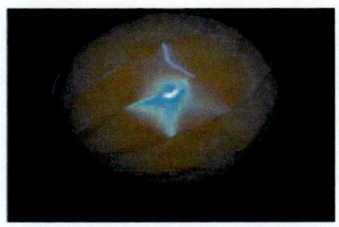

Figure 14. Photographs of cut piece of 1D PC in rectangular coordinates and PC in self-simulating spherical structure.

Now, a new technique is proposed: we first developed a technique to fabricate a heterostructure with gradual refractive index distribution and gradual period, and obtained 2D ODBG. Further, a self-simulating sphere structure was implemented based on this kind of heterostructure, and a genuine CBG in discretionary condition can be achieved.

Figure 15 shows the set-up geometry for fabricating our heterostructure. The laser used was a DPSS laser working at 457.9 nm with line width of 200 kHz (Melles Griot model 85-BLT-605). The holographic recording material was dichromated gelatin (DCG). The DCG emulsion of 36 μm thickness was coated on an optical glass plate with flatness of $\lambda/10$ and without any doping. The plate was mounted on a 1D translation stage driven by a stepping motor with a precision of 0.05 μm/step.

For an elementary reflection hologram made by two beams with 180°angle, because of the absorption by the recording material the distribution of refractive index inside the medium is a hyperbolic cosine function [30] (see Figure 16), but neither strictly sinusoidal nor strictly period. Besides, swelling of the DCG emulsion exists during the processing after exposure, and the swelling is not uniform [31]. So, we introduced a linear expansion parameter M_z, and the thickness swelling of jth layer at position z can be expressed as [31]

$$M_{z,j} = M_1 - (M_1 - M_N)j/N + r_j/a \tag{1}$$

where $M_{z,j}$ is the expansion parameter for jth period, r_j is a random number in the range [0,1], N is the number of layers, parameter a controls the amplitude of the random fluctuation. The linear random expansion is shown in Figure 17 [31]. So, the final result actually likes a chirped grating with a variable period.

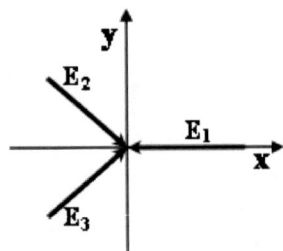

Figure 15. Set-up geometry of triangular lattice. The angles between E1 and E2, E2 and E3, E1 and E3 are all 120°.

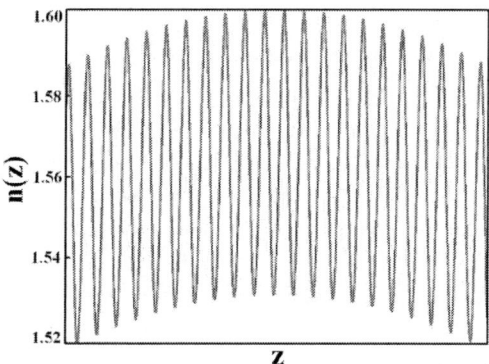

Figure 16. Schematic diagram of distribution of refractive index inside an element reflection hologram. T is the thickness of the material.

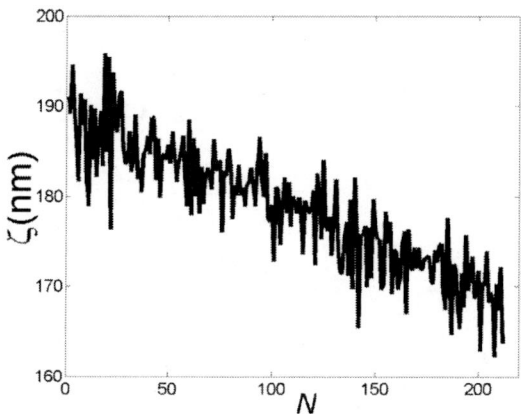

Figure 17. Linear random expansion series (Ordinate) of holographic recording material during the processing after exposure. Abscissa represents period numbers.

The results in Figures 16 and 17 show that what we obtained for an elementary reflection hologram is actually a heterostructure, and the interfaces of the heterostructure are gradual surfaces rather than step functions. In Figure 16, T is the thickness of the material. Also, the period is gradually changed rather than a constant. Therefore, the PC made by the method shown in Figure 15 is a PC heterostructure.

In our experiments, two exposures were taken for fabricating triangular lattices using the set-up geometry shown in Figure 15. After the first exposure, the recording material was translated a distance of d/3 along the $\Gamma - K$ direction of the first Brillouin zone, and then the second exposure was made. In this way, the lattice-point shape is not circular, but elongated and became elliptical. Figure 18 shows the transmission spectra of the triangular lattice heterostructure at different incident angles. It can be seen clearly that there is an absolute band gap in the wavelength range of 430 nm-500 nm at different incident angles, the width (FWHM) of the gap is about 50 nm (445-495 nm). It is an ODBG in deed. The oscillation at the bottom of the band gap is induced by the nonconforming periodicity.

The above experimental results could be explained theoretically as follows.

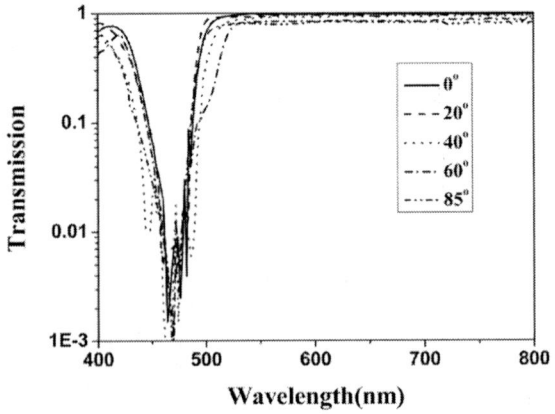

Figure 18. Transmission spectra of triangular lattice heterostructure made by two exposures with a translation.

For the dielectric columniation triangular lattice, the first band and the second band of TM wave at K of its first Brillouin zone is degenerated because of high symmetry. It results in that there is no common gap for TE and TM waves in $\Gamma - K$ direction. In our experiments, the method of two exposures with a translation was used, it made the lattice point shape to be non-circular (see the left iconograph at the bottom in Figure 19). So, the symmetry point group of this kind of triangular lattice is debased to C_{2v} from C_{6v}. It releases the band degeneration of TM wave at K of the first Brillouin zone, and this property is shown clearly in Figure 19 (pointed by the black arrows 1-4). The non-circular lattice-point shape of the triangular lattice induces two extra symmetric points K1 and M1 (see the right iconograph at the bottom in Figure 19)

The gradual period of the heterostructure is an important factor for obtaining ODBG. In this kind of structure the lattice constant is changed gradually and linearly. According to our experimental conditions, in our fabricated structure, there are 150 layers, they can be treated as 150 sub-structures, the minimum lattice constant is d=234.8 nm. During the processing of DCG after exposure a few air holes may be formed by dehydration, considering the refractive index of DCG be 1.52 before exposure, we choose the minimum refractive index $n_0 = 1$. The distribution of refractive index is proportional to light intensity and can be expressed by

$$n = n_0 + \frac{I}{I_{max} - I_{min}} \Delta n \tag{2}$$

$$I = \sum_{s=0, a/3} E_0^2 \left[1 + 4\cos\left(\frac{\sqrt{3}}{2}ky\right)\cos\left(\frac{3}{2}k(x+s)\right) + 4\cos^2\left(\frac{\sqrt{3}}{2}ky\right) \right] \tag{3}$$

where, s is displacement quantity along $\Gamma - K$ direction of the first Brillouin zone. The permittivity is

$$\varepsilon = n_0^2 + 2n_0 \Delta n \cdot \frac{I}{I_{max} - I_{min}} + \left(\frac{I}{I_{max} - I_{min}} \Delta n\right)^2 \tag{4}$$

where, $\Delta n = 0.5$, $I_{max} - I_{min} = 9$.

Because the heterostructure is a gradual period structure consist of 150 sub-structures, the upper band edge of the first sub-structure and the bottom band edge of the last (N=150) sub-structure determine the width of the total band gap of the heterostructure. In our calculations, the structure with the minimum period d (the first sub-structure) and that with the maximum period 1.15 d (the Nth sub-structure) were taken for calculations so that the upper and the bottom band edges can be obtained. Figure 19 gives the results calculated by plane wave expansion method. The dark solid line and the dark dots represent the bands of the first sub-structure for TM and TE waves respectively. The upper edges of the two bands form the high frequency edges of the heterostructure for TM and TE waves. The light dashed line and the light dots represent the bands of the Nth sub-structure for TM and TE waves respectively, The bottom edges of the two bands form the low frequency edges of the heterostructure for TM and TE waves. The total band gap of the heterostructure is the overlap of the band gaps of the 150 sub-structures. In any high symmetric direction of the first Brillouin zone, the directional band gap of the two sub-structures form the upper and the bottom band edges of the total band gap of the heterostructure in this direction respectively (as shown in light grey zone). The dark grey zone is the common gaps of all directional gaps so that an ODBG of the heterostructure is formed. The calculated position of the band gap locates at $3.8 \times 10^{15} \sim 4.1 \times 10^{15}$ s^{-1} (460nm~496nm), they (and the width) are well consistent with the experimental results.

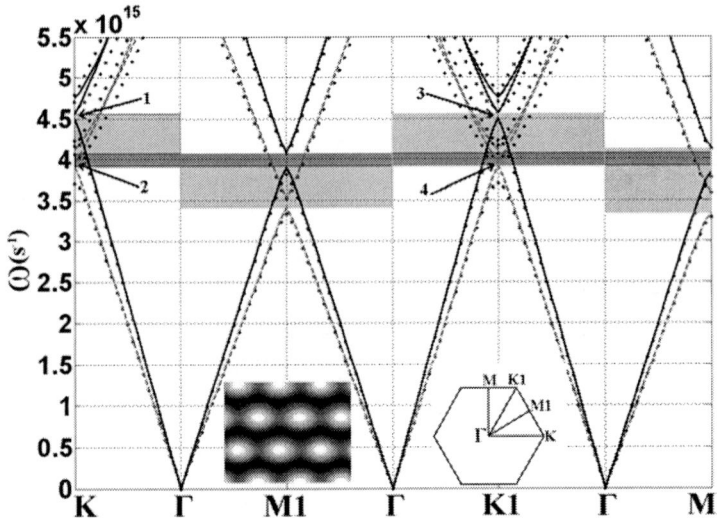

Figure 19. Calculated dispersion relation of gradual triangular lattice heterostructure. The left iconograph at the bottom shows the lattice geometry. The right iconograph at the bottom shows the first Brillouin zone of the triangular lattice.

One of the characters of this kind of triangular lattice is that the bands of TE and TM waves are almost superimposed, and along every direction band gaps are close to each other. This character is very helpful for achieving ODBG through gradual period heterostructures, when the layers of the heterostructure is more enough, and the expansion of the recording material reaches a certain range, a wider ODBG will be obtained by use of narrow directional band gaps. Obviously, it will be helpful for further works to achieve 3-D complete band gaps.

Based on the principle in Ref. [28] this kind of heterostructure can be transformed into a self-simulating spherical structure in the following steps: Firstly, preparing holographic recording material by coating 36 μm thick DCG emulsion on soft substrate polyethylene terephthalate (PET); Then, fabricating gradual index distribution and period heterostructure using the recording material; Finally, transforming the heterostructure into self-simulating sphere. This self-simulating sphere structure can be expressed by D_{nv} point group. Figure 20 shows a D_{nv} self-simulating sphere structure based on the gradual index distribution and gradual period heterostructure.

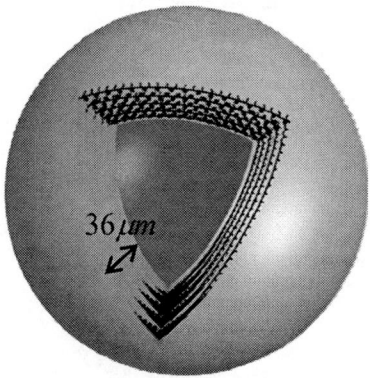

(a) Schematic diagram of D_{nV} self-simulating sphere structure.

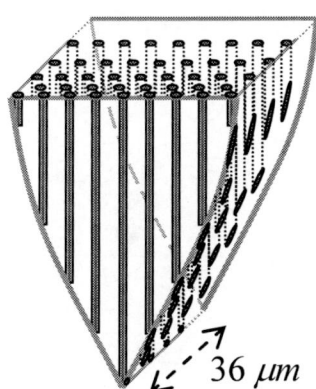

(b) The schematic diagram of the cut slice of the sphere shown in (a).

Figure 20. Structure of D_{nV} point group.

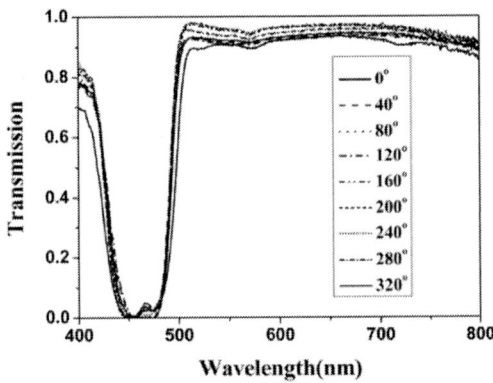

Figure 21. Measured transmission spectra of a D_{12V} self-simulating sphere structure for different value of φ at $\theta = 90°$.

The transmission spectra of a D_{12v} self-simulating sphere structure was measured and shown in Figure 21. In the measurements, a parallel beam with a diameter of 1 *mm* was used as the incident light, and was incident on the surface of the sphere passing through the center of the sphere. It is formed by a collimated beam passing through a 1 *mm* diaphragm. During the measurements, the angles θ and φ of the spherical system were changed from $0°$ through 180° and from 0° through 360° respectively. In our experiments, several D_{12v} self-simulating sphere structures with different diameters were fabricated. All of them showed identical characteristics under the same experimental conditions. It can be seen obviously that, for incident beams from different φ at $\theta = 90°$, the position of the band gap kept almost unchanged, and its width reaches about 70 nm.

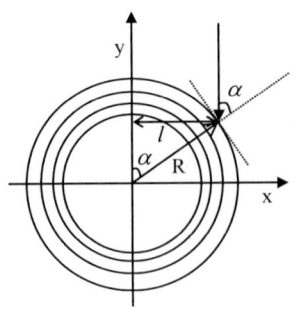

Figure 22. Schematic diagram when the incident light deviates from the center of the sphere. The corresponding incident angle is $\alpha = \arcsin(l/R)$.

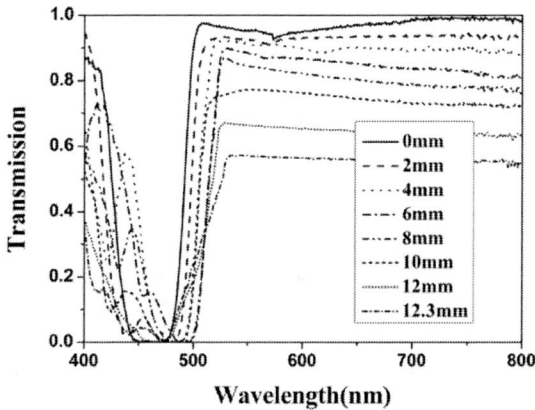

Figure 23. Measured transmission spectra of a D_{12v} self-simulating sphere structure for different values of φ at $\theta = 90°$ when the incident beam shifts different distances from the center. The radius of the sphere is 12.5 mm.

Besides, when the incident light is translated from the center of the sphere for different shifts l (see Figure 22), the transmission spectra are measured (see Figure 23). In the measurements, the radius of the sphere was 12.5 *mm*. It can be found that, even the shift is as large as 12.3 mm from the center of the sphere, there still is a CBG with a width of 40 nm. This shift corresponds to the condition that a light beam incidents on the surface of the sphere with an incident angle larger than $80°$. In our experiments, two D_{12v} self-simulating sphere structures were superimposed orthogonally to improve further the property of the band gap.

In summary, a D_{nv} point group structure based on 2-D gradual index distribution and gradual period heterostructure and self-simulating sphere possesses genuine complete band gap in a discretionary condition. The D_{nv} self-simulating sphere structure can be implemented easily by holography using low refractive index materials.

Chapter 5

THEORETICAL INVESTIGATION [41]

The band gaps of PC were described with various methods, such as plane wave expansion method (PWEM) [3,42,43], transfer matrix method (TMM) [44,45], multi-scattering theory [46], tight-binding formulation [47] and finite-difference-time-domain (FDTD) method [48,49]. However, only numerical solutions were obtained by these methods, in which their physical images are not intuitionistic or not clear enough. Therefore, a concise analytical solution (AS) will be valuable for analyzing more thoroughly the properties of the band gaps. Besides, these methods were mostly used for calculating the problems of PCs with step distribution of the refractive index. Recently, the present authors obtained very wide 3-D band gaps [23], even complete band gaps [28,33], using holographic technology and a low refractive index material basing on the principle of multi-beam interference [12,13], and has been used widely, only that the distribution of refractive index inside the recording material (such as photopolymers [50], liquid [51,52], or dichromated gelatin [23,28,33]) is gradual function rather than step function. Liu et al. [53] derived an analytical solution for 1-D PC with a sinusoidal distribution of refractive index. Samokhvalove et al. [54] made an approximation analysis of 2-D PC with rectangular dielectric rods. Nusinsky et al. [55] also made analytical calculation of rectangular PC. Analytical solutions presented by [54] and [55] are valid only for rectangular, step index photonic crystals. While the solution presented in this paper is valid for different structure having gradual distribution of refractive index.

Now, the theoretical analysis is investigated for the band gap characteristics of the PC made by multi beam interference. An analytical solution can be obtained

(1) BASIC CONSIDERATIONS

For multi-beam interference, the intensity distribution is a gradual function. When the distribution of the refractive index inside the recording material is proportional to the interference intensity, it should keep in consistency with the gradual distribution according to the property of interference. Therefore, only a few non-zero low order terms exist in the Fourier expansion of the dielectric constant. Since the wave function is generally a coupled multivariate linear equation group, a clear and simple AS can be obtained by approximation according to the spectral property.

For simplicity, an example of 2-D triangular lattice formed by interference of three coherent beams is shown in Figure 15. In Figure 15, the angle between and is , and points just to the bisector of this angle. Since polarization and phase have no effect on the final result, they were neglected in the following analysis.

The interference intensity in the x-y plane is

$$I = \left| \sum_{i=1}^{3} E_i \right|^2 = E_0^2 \left[1 + 4\cos(ky\sin\alpha)\cos[kx(1+\cos\alpha)] + 4\cos^2(ky\sin\alpha) \right] \quad (5)$$

Assuming that the permittivity of the material is proportional to the interference intensity, it will be $\varepsilon = n_0^2 + 2n_0 \Delta n \dfrac{I}{I_{max} - I_{min}}$ after exposure, here, n_0 is the minimum refractive index of the material, and Δn is the modulation of the refractive index. Its Fourier transform will be

$$F(\varepsilon) = \left(n_0^2 + 3C\right)\delta_{m,n} + C\left(\delta_{m-1,n+1} + \delta_{m-1,n} + \delta_{m,n+1} + \delta_{m,n-1} + \delta_{m+1,n} + \delta_{m+1,n-1}\right) \quad (6)$$

where,

$$C = \frac{2n_0 \Delta n}{I_{max} - I_{min}}.$$

It can be seen from Eq. (6) that there are only direct current (DC) term and first order term. Obviously, the value of DC term $n_0^2 + 3C$ is much larger (more than 20 times) than that of the first order term. For example, in dichromated gelatin (DCG) [23,28,33], where $n_0 = 1.52$, $\Delta n = 0.07$, for the triangular lattice formed by the inference of three beams with unit amplitude, we have $I_{max} - I_{min} = 9$, i.e., $\dfrac{n_0^2 + 3C}{C} \approx 101$, and all the first order terms are equal to C. So, the coupling among the first order terms could be neglected, and consider only the coupling between the DC term and any one of the first order term. So $\varepsilon_{\vec{G}'-\vec{G}}$ can be taken as $\begin{bmatrix} n_0^2 + 3C & C \\ C & n_0^2 + 3C \end{bmatrix}$. For DCG, the relation $n_0 + 3C \gg C$ could be satisfied when $\Delta n \leq 0.4$. All the band gaps belong to first Brillouin zone can be obtained by solving a coupled binary linear equation group, and each band gap corresponds to a simple AS.

(2) ANALYTICAL APPROACH (AS)

From Maxwell equations and Bloch'law, for TM wave, we have

$$det\left\{\left(abs\left(\vec{k}+\vec{G}'\right) abs\left(\vec{k}+\vec{G}\right)\right)/\varepsilon_{\vec{G}'-\vec{G}}\right\} = 0 \tag{7}$$

for TE wave, we have

$$det\left\{\left(\mathrm{Re}\left(\vec{k}+\vec{G}'\right)\mathrm{Re}\left(\vec{k}+\vec{G}\right) + \mathrm{Im}\left(\vec{k}+\vec{G}'\right)\mathrm{Im}\left(\vec{k}+\vec{G}\right)\right)/\varepsilon_{\vec{G}'-\vec{G}}\right\} = 0 \tag{8}$$

Figure 24 shows schematically a reciprocal lattice of a triangular lattice. The solid line represents the first Brillouin zone, and the dashed line represents the cell in the reciprocal lattice. K_1 and K_2 are two highly symmetric directions of the first Brillouin zone. G_1-G_6 are the representative reciprocal lattice vectors that collect all values of $\vec{G}$ and $\vec{G}'$ in Eqs. (7) and (8) in the cell of the reciprocal lattice. It can be seen that from the geometric relationship in Figure 2, $\left|\vec{k}+\vec{G}_0\right| = \left|\vec{k}+\vec{G}_1\right|$, where, $\vec{k}$ is in K_1 and K_2 direction respectively.

Let a be the lattice constant, so, the vector of its reciprocal lattice are b_1 and b_2 (see Figure 24).

For a triangular lattice

$$b_1 = \frac{2\pi}{a}\left(\hat{e}_x + \tan\left(\frac{\pi}{6}\right)\hat{e}_y\right), \quad b_2 = \frac{2\pi}{a\cos(\pi/6)}\hat{e}_y,$$

then we have

$$K_1 = \frac{4\pi}{3a}\hat{e}_x, \quad K_2 = \frac{\pi}{a}\hat{e}_x + \frac{\pi}{\sqrt{3}a}\hat{e}_y.$$

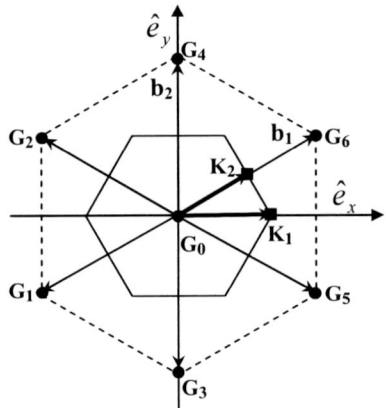

Figure 24. Schematic diagram of the reciprocal lattice of a triangular lattice. The dashed line represents the cell in the reciprocal lattice. The solid line represents the first Brillouin zone.

For TM wave, Eq. (7) can be transformed into

$$\begin{vmatrix} \frac{|\vec{k}+\vec{G}_0|^2(n_0^2+3C)}{(n_0^2+2C)(n_0^2+4C)} - \frac{\omega^2}{c^2} & \frac{-|\vec{k}+\vec{G}_0||\vec{k}+\vec{G}_i|C}{(n_0^2+2C)(n_0^2+4C)} \\ \frac{-|\vec{k}+\vec{G}_0||\vec{k}+\vec{G}_i|C}{(n_0^2+2C)(n_0^2+4C)} & \frac{|\vec{k}+\vec{G}_0|^2(n_0^2+3C)}{(n_0^2+2C)(n_0^2+4C)} - \frac{\omega^2}{c^2} \end{vmatrix} = 0$$

(9)

The positions of the top and the bottom edges of the first band gap are at

$$\omega_{\pm} = \sqrt{\frac{(n_0^2+3C)\pm C}{(n_0^2+2C)(n_0^2+4C)}}\left|\vec{k}+\vec{G}_0\right|c \tag{10}$$

So, $\omega_+ \approx \frac{1}{n_0}\left(1-\frac{2\Delta n}{9n_0}\right)|\vec{k}|c$ and $\omega_- \approx \frac{1}{n_0}\left(1-\frac{4\Delta n}{9n_0}\right)|\vec{k}|c$ can be obtained from Eq.(10) by using the condition $n_0^2+3C \gg C$ (i.e. $\Delta n \leq \frac{9}{34}n_0$).

Therefore, the following conclusions can be obtained:

1) The top and the bottom edges of the first band gap of TM wave are proportional to Δn and $|\vec{k}|$ ($|\vec{k}| \propto \frac{1}{\lambda}$, λ is the light speed in the material), and are inversely proportional to n_0. This is more definite than the common knowledge about "location of the band gap ($\frac{\omega_{l+}+\omega_{l-}}{2}$) is negatively related to the average refractive index", and shows the quantitative dependence of the top and the bottom band edges on material parameters.

2) For a certain material (i.e., n_0 and Δn are fixed), the band gaps will be determined only by the value of $|\vec{k}|$. So, the closer the shape of the first Brillouin zone approaches a circle, the smaller the variation of $|\vec{k}|$ (i.e., $|k_1| \approx |k_2|$). In this case, positions of the directional band gaps in K_1 and K_2 are almost same, obviously, it is more favorable for the appearance of absolute band gaps.

3) The relationship that $\frac{\Delta\omega}{\omega} = 2\frac{\omega_{l+}-\omega_{l-}}{\omega_{l+}+\omega_{l-}} \approx \frac{2\Delta n}{9n_0}|\vec{k}|c$ can be achieved (here $\Delta n \leq \frac{9}{34}n_0$), i.e. the bandwidth of the first band gap of TM wave is positively related to Δn and $|\vec{k}|$, and negatively related to n_0. This is more definite than the common knowledge "the higher the ratio of refractive indexes, the wider the band gaps".

Figure 25 shows the relationship between the band width $\Delta\omega/\omega$ and n_0 and Δn. In Figure 25, the color bar gives the normalized bandwidth $\Delta\omega/\omega$, the top line represents the width corresponding to the critical condition $\Delta n = (9/34)n_0$. It shows significantly that one does not have to seek for high refractive index materials for obtaining wider band gap. Band gap could readily be obtained with lower refractive index materials as same as that obtained with higher refractive index material (under the prerequisite condition of $\Delta n = (9/34)n_0$). Large value of the modulation of refractive index needs high refractive index, it is not definitely helpful for obtaining a broad band gap.

These features are obviously different from the relationship between the characteristics of band gap and the material parameters of 1-D PC [56]. But, the basic tendencies of variation of the band gap with material parameter obtained from both considerations are similar.

Also, for TE wave, Eq. (8) could be transformed into

$$\begin{vmatrix} \dfrac{\left|\vec{k}+\vec{G}_0\right|^2\left(n_0^2+3C\right)}{\left(n_0^2+2C\right)\left(n_0^2+4C\right)}-\dfrac{\omega^2}{c^2} & \dfrac{\mathrm{Re}\left[\left(\vec{k}+\vec{G}_0\right)\left(\vec{k}+\vec{G}_l\right)^*\right]C}{-\left(n_0^2+2C\right)\left(n_0^2+4C\right)} \\ \dfrac{\mathrm{Re}\left[\left(\vec{k}+\vec{G}_0\right)\left(\vec{k}+\vec{G}_l\right)^*\right]C}{-\left(n_0^2+2C\right)\left(n_0^2+4C\right)} & \dfrac{\left|\vec{k}+\vec{G}_0\right|^2\left(n_0^2+3C\right)}{\left(n_0^2+2C\right)\left(n_0^2+4C\right)}-\dfrac{\omega^2}{c^2} \end{vmatrix} = 0 \qquad (11)$$

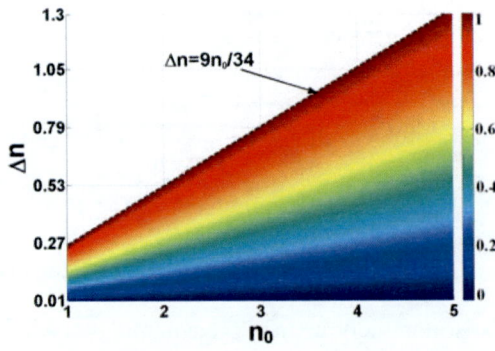

Figure 25. Calculated relationship between band width $\Delta\omega/\omega$ and n_0 and Δn. The color bar gives the band width $\Delta\omega/\omega$, the black dash line represents the width corresponding to $\Delta n = (9/34)n_0$.

The top and the bottom edges of the first band gap are at

$$\omega_{\pm} = c\sqrt{\frac{\left|\vec{k}+\vec{G}_0\right|^2\left(n_0^2+3C\right)\mp \mathrm{Re}\left[\left(\vec{k}+\vec{G}_0\right)\left(\vec{k}+\vec{G}_1\right)^*\right]C}{\left(n_0^2+2C\right)\left(n_0^2+4C\right)}} \tag{12}$$

Let β be the angle between K_1 and K_2 in the direction of K_1 (see Figure 24), then, $\omega_{i\pm} = \sqrt{\frac{\left(n_0^2+3C\right)\pm C\cos 2\beta}{\left(n_0^2+2C\right)\left(n_0^2+4C\right)}}\left|k+G_0\right|c$ which is consistent with Eq.(10). It shows that the influencing factors on the width of the first band gap of TE wave, and the rule of their variations are as same as those for TM wave.

In K_2 direction, $\left(\vec{k}+\vec{G}_1\right)^* = -\left(\vec{k}+\vec{G}_0\right)^*$, then Eq. (8) could be transformed to Eq. (6). This is the reason why same results for TE and TM waves are always obtained in the direction of K_2 calculated by PWEM under the condition of small modulation of refractive index.

(3) COMPARISON WITH PLANE WAVE EXPANSION METHOD (PWEM)

It is clear from Eqs. (9) and (11) that the eigenvalue ω^2/c^2 is positively related to the length of $\vec{k}+\vec{G}$. From Figure 24, it can be seen that the lengths of $\left|\vec{k}+\vec{G}_0\right|$ and $\left|\vec{k}+\vec{G}_1\right|$ are minimal in both K_1 and K_2 directions. They correspond respectively to the minimal eigenvalue ω^2/c^2, i.e. the lowest band gaps in two highest symmetric directions.

Figure 26 gives the calculated results using Eqs. (10) and (12). It shows that, $\left|\vec{k}+\vec{G}_0\right|$ and $\left|\vec{k}+\vec{G}_1\right|$ determine the top and the bottom limit of the first gap, respectively. Each direction of G in Figure 24 corresponds to two band gaps (one for TM wave and the other for TE wave). The top and the bottom edges of the gaps corresponding to TM and TE waves can be obtained by solving Eqs. (9) and (11) when the collection of the values of $\vec{G}$ and $\vec{G}'$ is defined as $\{\vec{G}_0,\vec{G}_i\}$ $(i=1,2,3,4,5,6)$. The value at any point of the band can be calculated by reducing the value of K while keeping its direction unchanged.

The inserts in Figure 4 pointed by the arrows are the zoomed figures of the first band gap in K_1 and K_2 directions respectively, in which the position and the width obtained by PWEM are in good consistency with those obtained by our AS. The bandwidth of TM wave obtained by AS is slightly wider than that obtained by PWEM in K_1 direction, while the width of the gaps of TE wave is slightly narrower. In K_2 direction, those obtained by the two methods coincide almost perfectly. However, it should be noted that in K_1 direction there is no gap for TM wave according to PWEM, but, in AS which is not zero. It indicts that this approximation is more accurate to predict the location of band gaps than the bandwidth of band gaps for TM polarization.

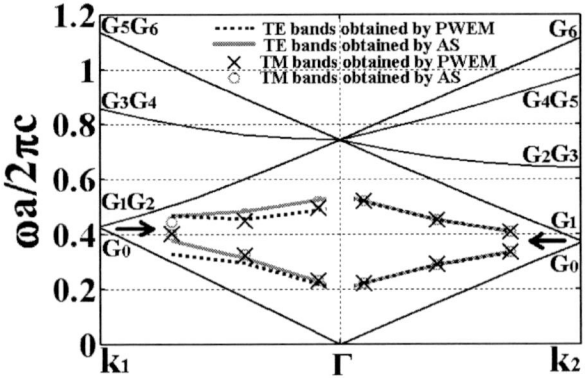

Figure 26. Band gaps calculated by AS and PWEM.

For detail comparison between the two methods, the variation of the top and the bottom band edges with Δn was obtained (see Figure 27) by analyzing Eqs. (10) and (12).

Analytical Solution

In Figure 27, it can be seen that: the top and the bottom edges of TM and TE waves in K_2 direction are in good superposition. In other words, the dielectric constants can be considered as the same for TE and TM waves in K_2 direction when Δn is small. The bandwidth in K_1 and K_2 directions increase with the increase of the refractive index modulation. The space between the bottom edge in K_1 direction and the top edge in K_2 direction decreases with the increase of the refractive index modulation, but they won't intercept when $\Delta n \leq 0.4$. It means that there is no absolute band gap in low refractive index materials with strict period structure for triangular lattice. But, it shows that the bottom edge in K_1 direction and the top edge in K_2 direction will superimpose, and absolute band gap will appear when Δn is large enough. Or, in another way of thinking, absolute band gap could be obtained by optimizing the lattice structure, i.e. using a non-strictly periodical structure under low refractive index modulation[28,33].

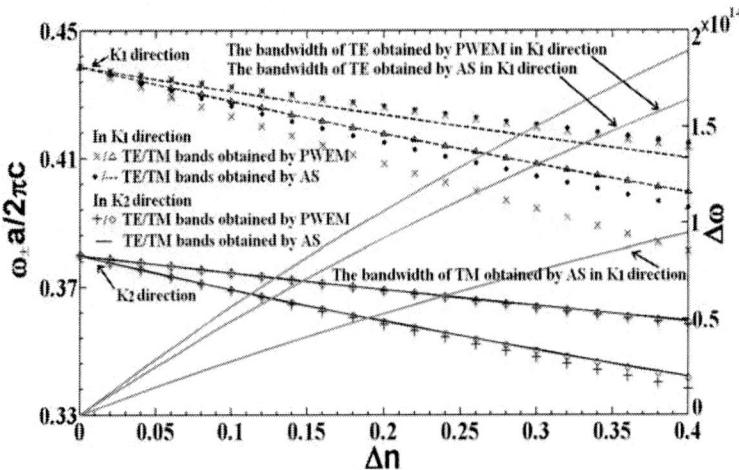

Figure 27. Variation of Position of band edges and bandwidth with modulation of refractive index calculated by AS and PWEM.

Plane Wave Expansion Method

In Figure 27, it can be seen that: the TM mode degenerates in K_1 direction. The top edge of the TE band and the bottom edge of the TM band in K1 direction are in good consistency with those obtained by AS, but, the bandwidth is different from that obtained by AS. While, the band location and the bandwidth of TM and TE waves in K_2 direction obtained by the two methods are all well consistent, and show almost no difference when $\Delta n \leq 0.1$.

(4) CONCLUSION

Analytical solution was obtained for 2D PCs made by holography. It gives concise and intuitionistic physical image as well as the relationship between the characteristics of the band gaps and the parameters of the materials. The results obtained by the analytical solution are in good consistency with those obtained by plane wave expansion method, but, it should be noted that in some directions the solution may not accurate enough and may be different from the degenerate case in which predicts by PWEM. This analytical approach is practically the first order approximation of the solution of the PC with step distribution of refractive index. Though the analytical solution is derived for 2D triangular lattice made with a special material (DCG), it is valid for any other kind of 2D lattice made by multi-beam interference or other gradual refractive index materials. And it must be noted that this solution is accurate only for low refractive index modulation materials.

Chapter 6

TEMPERATURE TUNABLE RANDOM LASING IN WEAKLY SCATTERING STRUCTURE FORMED BY SPECKLE

Recently, the low refractive index material DCG was used to implement a weakly scattering disorder structure formed by speckle, and temperature tunable random lasing using this kind of structure was achieved.

Since the pioneering work of Ambartsumyan et al. [57], random laser has been a subject of intense theoretical and experimental studies [58-65]. In particular, the discovery of intriguingly narrow spectral emission peaks (spikes) in certain random lasers by Cao et al. [60] has given an enormous boost to such a subject. The random laser represents a nonconventional laser whose feedback is mediated by random fluctuations of the dielectric constant in space. In general, there are two kinds of feedback for the random lasing, intensity feedback and field feedback [64,65]. The former is phase insensitive, which is called as incoherent or nonresonant feedback. The latter is phase sensitive, which can be regarded as coherent or resonant feedback.

The two types of random laser action can not only be found in some strong scattering systems with gain [64-67], they can also be observed in active random media in weakly scattering regime far from Anderson localization. For example, some authors have reported laserlike emission from several weakly scattering samples including conjugated polymer films, semiconductor powders and dye-infiltrated opals [68-72]. In the strongly scattering regime, lasing modes have a nearly one-to-one correspondence with the localized modes of the passive system [64-66,73]. In contrast, the nature of lasing modes in weakly scattering open random systems is still under discussion, although several mechanisms have been proposed to explain such a phenomenon [74-77].

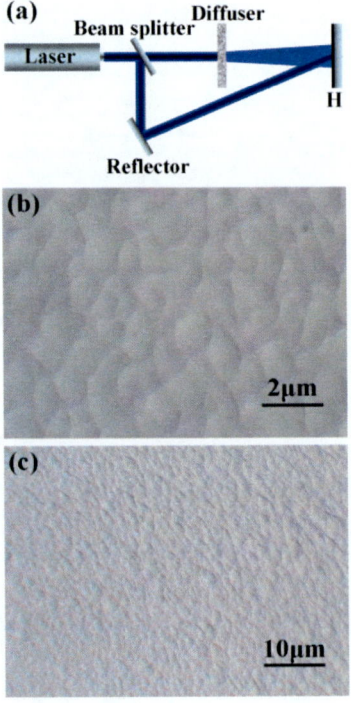

Figure 28. (a) Set-up geometry of the disorder structure formed by speckle. H is holographic recording material. (b) and (c) are the microscopic images of the disordered structure formed by speckle.

Here we report an experimental observation of random laser in a kind of new random medium in a regime of weak scattering. A disorder structure was formed by speckle. The set-up geometry is shown in Figure 28 (a) schematically. A ground glass was used to generate spatial speckle structure. A reference beam was introduced to implement interference for recording the distribution of the spatial speckle in the holographic recording material. Different distribution of the spatial speckle can be recorded by changing the place of the ground glass or the reference beam. The recording material used was dichromated gelatin (DCG) coated on optical glass. The thickness of the DCG used is $36\mu m$. After exposure, the DCG plate was developed in running water for 120 min at 20°C for developing sufficiently and to remove any residual dichromate, and then soaked into a Rhodamine 6G solution with a concentration of 0.125 mg per milliliter water at the same temperature for 60 min bath enabling the dye molecules to diffuse deep into the emulsion of the gelatin. Then, the DCG plate was dehydrated in turn by soaking it in 50 %, 75 % and 100 % isopropyl alcohol containing the Rhodamine 6G dye with same concentration at 40°C for 15 min.

After dehydration, the DCG plate was baked at 100°C for 60 min in an oven. Figure 28 (b) and (c) show the microscopic image of the disordered structure formed by the speckle, and it is a complete disordered structure. The average size of the speckle spot is about $1\,\mu m$. The concentration of the speckle spot is estimated to be the order of magnitude of 10^9 per cm^3. The DCG is a phase type holographic recording material with the refractive index of 1.52 and the modulation of the refractive index less than 0.1. From the transmission measurements, we can estimate the mean free path $l^* > 80\,\mu m$ in the spectral range between 550 nm and 600 nm, which is much larger than the thickness of the system. This means that scattering in the structure is quite weak.

We now turn to the investigations on the photoluminescence (PL) of the above dye-doped random samples under a pump field. Figure 29 (a) shows the set-up geometry for the measuring. A nanosecond (ns) pulsed Nd:YAG laser running at 532 nm with repetition rate of 10 Hz, maximum pulse energy of 1500 mJ , pulse width of 8 ns and a picosecond (ps) pulsed Nd:YAG laser running at 532 nm with repetition rate of 10 Hz, maximum pulse energy of 40 mJ , pulse width of 30 ps were used as the pumping sources respectively. The beam diameters of the ns laser and the ps laser were 13 mm and 3 mm respectively. The laser beam was incident on the surface of the optical glass substrate, then penetrates the substrate and entered the dye doped DCG medium. The detector was put close to the sample to collect the energy as much as possible. By rotating the rotation stage the orientation of the speckle structure can be changed with respect to the laser beam. It results in the variation of the wavelength and the peak number of lasing.

The measured results for the emission spectra under the pump by ps laser are plotted in Figure 29 (b). The blue triangles correspond to the PL spectra of Rhodamine 6G, and the red solid line represents the emission of the dye-doped sample at $\theta = 25°$ under the pumping intensity of 120 MW/mm^2. The threshold of the random laser is about 50 MW/mm^2 as shown in the insets. It can be seen that many lasing modes are excited in the gain spectrum of Rhodamine 6G. The widths of these modes are found to be less than 0.4 nm, which exhibits the coherent feedback very well. The phenomenon is very similar to the previous investigations on the random laser in the other weak scattering systems [68-72]. The origin of the phenomenon can also be understood by the previous theory for such a problem [74-77].

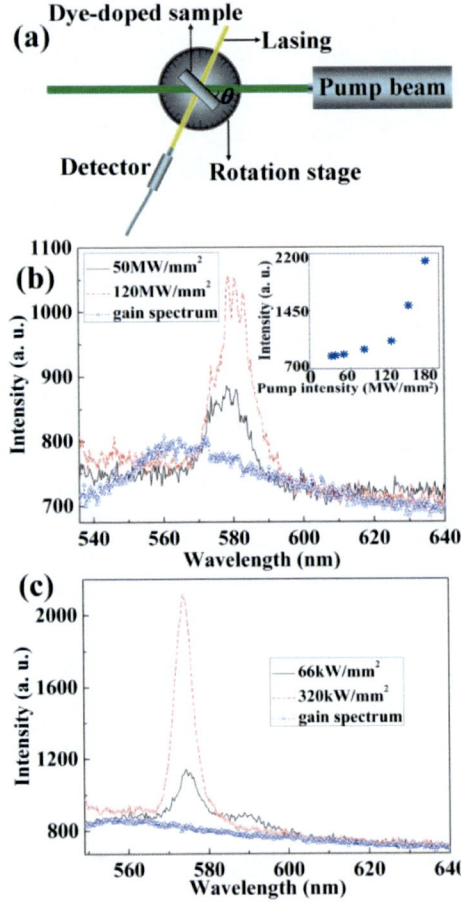

Figure 29. (a) Set-up geometry for measuring random lasing. θ is the angle of the pump beam with respect to the surface of the substrate, it represents the orientation of the sample. (b) Emission pumped by ps laser, and the inset shows the threshold is about 50MW/mm^2 at $\theta = 25°$. (c) Emision pumped by ns laser at $\theta = 30°$. In (b) and (c), the blue triangles are the gain spectrum of rhodamine 6G excited by 532 nm laser beam. The black solid line and the red dashed line represent the emissions with pumped energy around and much higher than threshold respectively.

However, when the dye-doped random sample is pumped by the ns laser, the different phenomenon appears. Figure 29(c) displays the measured results when the pump energy is much higher (320 kW/mm^2) than and around (66 kW/mm^2) the threshold. A single mode is observed and the profile of the mode is fitted by a Gaussian function, and the result demonstrates that this mode shows a perfect Gaussian profile (as shown in Figure 30(a)). Such a character does not change with the change of the pump energy. When $\theta = 30°$, the emission spectra at different pump energy are shown in Figure 30(b). There is only a single mode at 574 nm. The threshold value is found to be 37.7 kW/mm^2, which is much lower than that pumped by ps laser (50 MW/mm^2). The lasing becomes more obvious when the pump energy is higher than 65.9 kW/mm^2.

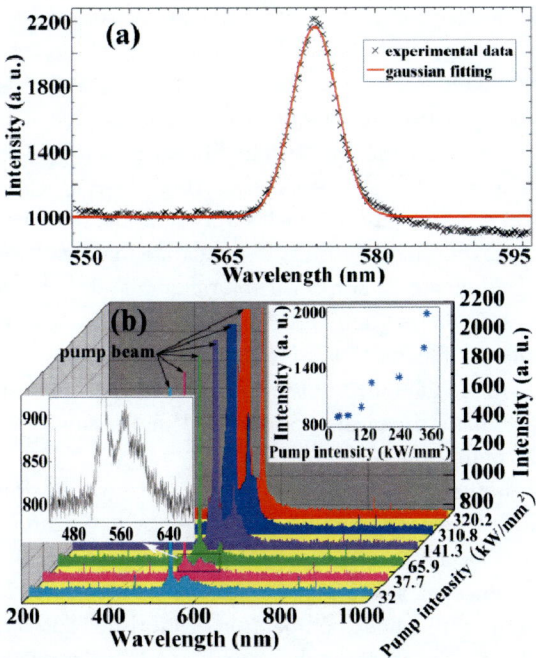

Figure 30. (a) Gaussian function fitted emission peak. (b) Measured emission spectra at different pump energy in the case of $\theta = 30°$, and the inset shows the threshold is about 37.7 kW/mm^2. The left high spikes (marked by black arrows) correspond to 532 nm pump laser beam, and the appeared width does not show its actual line width of the pump laser because of the saturation effect of the CCD detector used.

In fact, our sample is in the shape of a thin film, it can not be considered as a strict isotropic structure. So, the property of the mode also depends on the pump direction of the laser beam. In some directions, the case with two modes can also be found. Figure 31(a) shows the measured results for the emission spectra of the disordered structure at different pump energy using the ns laser. There are two modes of the emission located at 570 nm and 590 nm respectively and the later is much stronger than the former, and the later becomes the dominant mode when the pump energy is strong enough. When the pump energy is around the threshold value, the two modes are in competition (see Figure 31(b)).

In addition, we find that the widths of the modes always keep a few nm for any angle (direction) and intensity of the ns pump beam, which displays the character of intensity feedback. In general, the intensity feedback corresponds to diffusion motion of photons in active random medium, when the photon mean free path is much smaller than the dimension of scattering medium but much longer than the optical wavelength [64,65]. It can be described theoretically by the diffusion equation for the photon energy density in the presence of a uniform and linear gain [57,64,65]. Obviously, it is not such a case for the weak scattering system in the present work, because the mean free path is much larger than the thickness of the sample. However, our experimental results demonstrate that incoherent feedback can also be realized in such a non-diffusion random system by choosing suitable pump sources. In the previous investigations, one has observed the transition between coherent feedback and incoherent feedback by varying the amount of scattering in the gain medium [64,78]. In fact, our present work illustrates that such a transition can also be realized in the same random sample by changing the pump source.

The phenomena can be understood in terms of the following analysis. For the case of ns laser pumping, the light will go through a path about several meters in the period of one pulse. It means that the light will have much more times of scattering, i.e. has much longer effective interaction length. Therefore, the mode competition has finished and one mode becomes dominant. That is the reason only a single mode is dominant in the emission as long as the pumping energy is higher. However, for the case of ps laser pumping, the light will go through a path about several millimeters. It means that the effective interaction length is much shorter, so the mode competition can not be finished even though the pumping energy is rather high. Besides, for the case pumped by ps laser, because the effective interaction length is very short, i.e., the scattering times is much less. Therefore, much higher pump energy is needed for building up the lasing, it results in the high threshold.

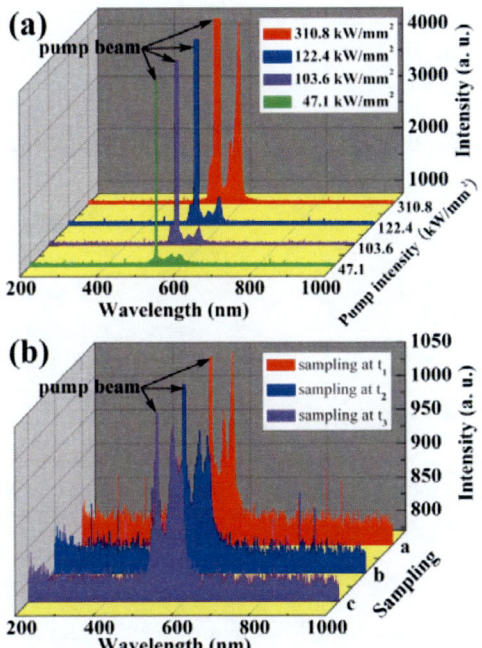

Figure 31. (a) Measured radiation spectra at different pump energy in the case of $\theta = 18°$. (b) Competition of two modes around the threshold value in the case of $\theta = 18°$. The left high spikes (marked by black arrows) correspond to 532 nm pump laser beam, and the appeared width does not show its actual line width of the pump laser because of the saturation effect of the CCD detector used.

Another interesting phenomenon for the present system is that the emission wavelength for the random laser can be tuned through changing temperature. Figure 32 shows the measured spectra of lasing emission at different temperatures. It can be seen clearly that the emission peaks shift toward to the long wavelength with the increase of the temperature. The lasing will be kept so long as the frequency of the emission is still inside the gain profile when temperature changes. In fact, the temperature tuning for the random laser by infiltrating sintered glass with laser dye dissolved in a liquid crystal had been investigated in previous work [62]. The diffusive feedback was controlled through a change of refractive index of the liquid crystal with temperature. However, such a tuning was employed to turn on and off random lasers. In contrast, the wavelength tuning can be realized in the present system. The phenomenon originates from the special property of the material. With the changes of the temperature, the light path changes due to variation of the distance between the two adjacent speckle spots. It makes the variation of the effective length of interaction, and then the lasing wavelength is changed.

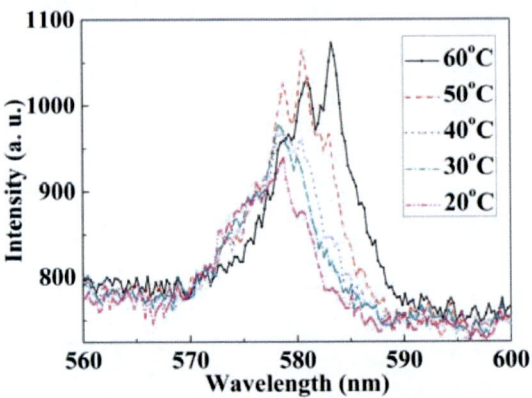

Figure 32. Tunability of wavelength vs. temperature of the random laser.

Finally, it should be pointed out that, the DCG can be coated on a soft substrate, such as polyethylene terephthalate (PET). So, the speckle structure recorded by holography can be formed as a soft film. On the other hand, the holography recorded speckle structure is pretty stable. Hence, the output of this kind laser has good stability and repeatability. It will be beneficial for actual applications. We did implement the speckle random laser using DCG coated on the PET soft substrate, and same results mentioned above were obtained.

In summary, we have fabricated a weakly scattering disordered structure formed by speckle using holography. In such a system with gain, we have observed low-threshold random lasing with two kinds of feedback, incoherent and coherent, for the same sample in the case pumped by nanosecond and picosecond lasers, respectively. The wavelength tunability of the random laser with the change of the temperature has been demonstrated. Our results can not only get a deeper understanding on the open question about the properties of the lasing modes in weakly scattering regime, it can also benefit some applications such as remote temperature sensing in hostile environments.

REFERENCES

[1] John, S. "Strong localization of photons in certain disordered dielectric superlattices," *Phys. Rev. Lett.*, vol. 58, 1987, 2486-2489.
[2] Yablonovitch, E. "Inhibited spontaneous emission in solid-state physics and electronics," *Phys. Rev. Lett.*, vol. 58, 1987, 2059-2062.
[3] Ho, KM; Chan, CT; Soukoulis, CM. *Phys. Rev. Lett*, 1990, 65, 3152-3155.
[4] Chan, TYM; Toader, O; John, S. *Phys. Rev. E*, 2005, 71, 046605.
[5] Sharp, DN; Turberfield, AJ; & RG. Denning, *Phys. Rev. B*, 2003, 68, 205102.
[6] Meisel, DC; Wegener, M; Busch, K. *Phys. Rev. B*, 2004, 70, 165104.
[7] Blanco, A. Emmanuel Chomski, Serguei Grabtchak, Marta Ibisate, Sajeev John, Stephen W. Leonard, Cefe Lopez, Francisco Meseguer, Hernan Miguez, Jessica P. Mondia, Geoffrey A. Ozin, Ovidiu Toader and Henry M. van Driel, *Nature*, 2000, 405, 437-440.
[8] Vlasov, YA; Bo, XZ; Sturm, JC; Norris, DJ. *Nature*, 414, 289-293.
[9] Wu, LJ; Wong, KS. *Appl. Phys. Lett*, 2005, 86, 241102.
[10] Lin, SY ; Fleming, JG; Hetherington, DL; Smith, BK; Biswas, R; Ho, KM; Sigalas, MM; Zubrzycki, W; Kurtz, SR; Jim Bur, *Nature*, 1998, 394, 251- 253.
[11] Noda, S; Tomoda, K; Yamamoto, N; Chutinan, *Science*, 2000, 289, 604-606.
[12] Berger, V; Gauthier-Lafaye, O; Costard, E. *J. Appl. Phys.*, 1997, 82, 60-64.
[13] Campbell, M; Sharp, DN; Harrison, MT; RG. Denning, *Nature*, 2000, 404, 53-56.

[14] Zhong, YC ; Zhu, SA; Su, HM; Wang, HZ; Chen, JM; Zeng, ZH; Chen, YL. *Appl. Phys. Lett*, 2005, 87, 061103.
[15] Toader, TYM. Chan and S. John, *Appl. Phys. Lett*, 2006, 89, 101117.
[16] Kondo, T; Matsuo, S; Juodkazis, S. Misawa., *Appl. Phys. Lett*, 2001, 79, 725-727.
[17] Miklyaev, YV; Meisel, DC; Blanco, A; Freymann, GV; Busch, K; Koch, W; Enkrich, C; Deubel, M; Wegener, M. *Appl. Phys. Lett.*, 2003, 82, 1284-1286.
[18] Divliansky, TS; Mayer, KS; Holliday, VH. Crespi, *Appl. Phys. Lett*, 2003, 82, 1667-1669.
[19] Meisel, DC; Diem, M; Deubel, M; Willard, FP; Linden, S; Gerthsen, D; Busch, K; Wegener, M. *Adv. Mater.*, 2006, 18, 2964-2968.
[20] Cui, LB; Wang, F; Wang, J; Wang, ZN; Liu, DH. *Phys. Lett. A*, 2004, 324, 489-493.
[21] Zoorob, ME; Charlton, MDB; Parker, GJ; Baumberg, JJ; Netti, MC. *Nature* (London), 2000, 404, 740.
[22] Zhang, X; Zhang, ZQ; Chan, CT. *Phys. Rev. B*, 2001, 63, 081105.
[23] Zhi Ren, Zhaona Wang, Tianrui Zhai, Hua Gao, Dahe Liu, Xiangdong Zhang, *Phys. Rev. B*, 2007, 76, 035120.
[24] Su, HM; Zhong, YC; Wang, X; Zheng, XG; Xu, JF; Wang., HZ. *Phys. Rev. E*, 2003, 67, 056619..
[25] Wang, X; Xu, JF; Su, HM; Zeng, ZH; Chen, YL; Wang, HZ; Pang, YK; Tam, WY. *Appl. Phys. Lett*, 2003, 82, 2212-2214.
[26] Dahe Liu, Weiguo Tang, Wunyun Huang, and Zhujian Liang,, *Opt.Eng.*, 1992, 31, 809-812.
[27] Smith ed., HM. *Holographic Recording Materials*, Ch.3, Springer-Verlag, Berlin, 1977.
[28] Zhi Ren, Tianrui Zhai, Zhaona Wang, Jing Zhou, Dahe Liu, *Advanced Materials*, 2008, 20, 2337-2340.
[29] Tianrui Zhai, Zhi Ren, Zhaona Wang, Jing Zhou, Dahe Liu, *IEEE Photon. Tech. Lett.*, 2008, 20, 1066-1068.
[30] Liu, D; Zhou, J. Opt. *Commun.* 1994, 107, 471.
[31] Wang, Z; Liu, D; Zhou, J. Opt. *Lett*, 2006, 31, 3270.
[32] Wang, X; Wang, F; Cui, L; Liu, D. Opt. *Commun.*, 2003, 221, 289.
[33] Tianrui Zhai, Zhaona Wang, Rongkuo Zhao, Jing Zhou, Dahe Liu, Xiangdong Zhang, *Appl. Phys. Lett.*, 2008, 93, 210902.
[34] Emanuel Istrate, Edward H. Sargent, *Rev. Mod. Phys.*, 2006, 78, 455-481.

[35] Jiang, P; Ostojic, GN; Narat, R; Mittleman, DM; Colvin, VL. Adv. Mater. (Weinheim, Ger), 2001, 13, 16.
[36] Egen, M; Voss, R; Griesebock, B; Zentel, R; Romanov, S; Torres, CMS. *Chem. Mater*, 2003, 15, 3786-3792.
[37] Wong, S; Kitaev, V; Ozin, GA. *J. Am. Chem. Soc.*, 2003, 125, 15589, 15598.
[38] Song, BS; Noda, S; Asano, T. *Science*, 2003, 300, 1537-1537.
[39] Srinivasan, K; Barclay, PE; Painter, O; Chen, J; Cho, AY; Gmachl, C. *Appl. Phys. Lett*, 2003, 83, 1915.
[40] Kawakami, S; Sato, T; Miura, K; Ohtera, Y; Kawashima, T; Ohkubo, H. *IEEE Photonics Technol. Lett*, 2003, 15, 816-818.
[41] Tianrui Zhai, Zhaona Wang, Rongkuo Zhao, Xiaobin Ren, and Dahe Liu, IEEE J. Quant. *Electron.*, 2009, 45, 1297-1301.
[42] Leung, KM; Liu, YF. "Full vector wave calculation of photonic band structures in face-centered-cubic dielectric media," *Phys. Rev. Lett.*, 1990, vol. 65, 2646-2649.
[43] Zhang, Z; Satpathy, S. "Electromagnetic wave propagation in periodic structures: Bloch wave solution of Maxwell's equations," *Phys. Rev. Lett.*, 1990, vol. 65, 2650-2653.
[44] Pendry, JB; MacKinnon, A. "Calculation of photon dispersion relations," *Phys. Rev. Lett.*, 1992, vol. 69, 2772-2775.
[45] Pendry, J. "Photonic band structures," *J. Mod. Opt.*, vol, 1994, 41, 209-229.
[46] Li, L; Zhang, Z. "Multiple-scattering approach to finite-sized photonic band-gap materials," *Phys. Rev. B*, 1998, vol. 58, 9587-9590.
[47] Lidorikis, E; Sigalas, M; Economou, E; Soukoulis, C. "Tight-binding parametrization for photonic band gap materials," *Phys. Rev. Lett.*, 1998, vol. 81, 1405-1408.
[48] Chan, CT; Datta, S; Yu, QL; Sigalas, M; Ho, KM; Soukoulis, CM. "New structures and algorithms for photonic band gaps," *Physica A*, 1994, vol. 211, 411-419.
[49] Tran, P. "Photonic-band-structure calculation of material possessing Kerr nonlinearity," *Phys. Rev. B*, 1995, vol. 52, 10673-10676.
[50] Jakubiak, R; Bunning, T; Vaia, R; Natarajan, L; Tondiglia, V. "Electrically switchable, one-dimensional polymeric resonators from holographic photopolymerization: A new approach for active photonic bandgap materials," *Adv. Mater.*, 2003, vol. 15, 241-244.
[51] Mach, P; Wiltzius, P; Megens, M; Weitz, D; Lin, K; Lubensky, T; Yodh, A. "Electro-optic response and switchable Bragg diffraction

for liquid crystals in colloid-templated materials," *Phys. Rev. E*, 2002, vol. 65, 31720-31723.
[52] Jakubiak, R; Tondiglia, VP; Natarajan, LV; Sutherland, RL; Lloyd, P; Bunning, TJ; Vaia, RA. "Dynamic lasing from all-organic two-dimensional photonic crystals," *Adv. Mater.*, 2005, vol. 17, 2807-2811.
[53] Zheng, J; Ye, Z; Wang, X; Liu, D. "Analytical solution for bandgap structures in photonic crystal with sinusoidal period," *Phys. Lett. A*, 2004, vol. 321, 120-126.
[54] Samokhvalova, K; Chen, C; Qian, B. "Analytical and numerical calculations of the dispersion characteristics of two-dimensional dielectric photonic band gap structures," *J. Appl. Phys.*, 2006, vol. 99, 063104.
[55] Nusinsky, I; Hardy, A. "Approximate analysis of two-dimensional photonic crystals with rectangular geometry. I. E polarization," *J. Opt. Soc. Am. B*, 2008, vol. 25, 1135-1143.
[56] Joannopoulos, JD; Johnson, SG; Winn, JN; Meade, RD. *Photonic Crystals: Molding the Flow of Light*, 2nd ed. Princeton, NJ:Princeton Univ. Press, 2008, 49-52.
[57] Ambartsumyan, RV; Basov, NG; Kryukov, PG; Letokhov, VS. IEEE J. Quantum Electron. QE-2, 442 (1966), V. S. Letokhov, Sov. *Phys. JETP*, 26, 835, 1968.
[58] Lawandy, N; Balachandran, R; Gomes, A; Sauvain, E. *Nature* (London), 368, 436 1994.
[59] Wiersma, D; van Albada, M; Lagendijk, A. *Phys. Rev. Lett*, 1995, 75, 1739.
[60] Cao, H; Zhao, YG; Ong, HC; Ho, ST; Dai, JY; Wu, JY; Chang, RPH. *Appl. Phys. Lett*, 73, 3656, 1998, *Phys. Rev. Lett*, 1999, 82, 2278.
[61] Jiang, XY; Soukoulis, CM. *Phys. Rev. Lett*, 2000, 85, 70.
[62] Wiersma, D; Cavalieri, S. *Nature* (London), 2001, 414, 708-709. A. Rose, Zhengguo Zhu, Conor F. Madigan, Timothy M. Swager and Vladimir Bulovic, *Nature* (London), 2005, 434, 876.
[63] Hanken, E. Türeci, Li Ge, Stefan Rotter, A.Douglas Stone, *Science*, 2008, 320, 643.
[64] Cao, H. *Waves Random Media*, 2003, 13, R1 2003.
[65] Wiersma, DS. *Nature Physics*, 2008, 4, 359.
[66] Milner, V; Genack, AZ. *Phys. Rev. Lett*, 2005, 94, 073901.
[67] Fallert, J; Dietz, R; Sartor, J; Schneider, D; Klingshirn, C; Kalt, H. *Nature Photon*, 2009, 3, 279.

[68] Frolov, SV; Vardeny, ZV; Yoshino, K; Zakhidov, A; Baughman, RH. *Phys. Rev., B*, 1999, 59, R5284.
[69] Ling, Y; Cao, H; Burin, AL; Ratner, MA; Liu, X; Chang, RPH. *Phys. Rev. A*, 2001, 64, 063808.
[70] Mujumdar, S; Ricci, M; Torre, R; Wiersma, D. *Phys. Rev. Lett*, 2004, 93, 053903.
[71] Polson, RC; Vardeny, ZV. *Phys. Rev. B*, 2005, 71, 045205.
[72] Wu, X; Fang, W; Yamilov, A; Chabanov, A; Asatryan, A; Botten, L; Cao, H. *Phys. Rev. A*, 2006, 74, 53812.
[73] Jiang, X; Soukoulis, CM. *Phys. Rev. E*, 2002, 65, 025601.
[74] Apalkov, VM; Raikh, ME; Shapiro, B. *Phys. Rev. Lett*, 2002, 89, 016802.
[75] Florescu, L; John, S. Phys. *Rev. Lett*, 2004, 93, 13602.
[76] Deych, L. *Phys. Rev. Lett*, 2005, 95, 043902.
[77] Vanneste, C; Sebbah, P; Cao, H. *Phys. Rev. Lett*, 2007, 98, 143902.
[78] Cao, H; Xu, JY; Chang, SH; Ho, ST. *Phys. Rev. E*, 2000, 61 1985.

INDEX

A

alcohol, 49
amplitude, 26, 37
anisotropy, 2
applications, 2, 8, 21, 25, 55, 56
authors, 1, 35, 47

B

band gap, 1, 2, 8, 11, 15, 17, 20, 21, 22, 23, 28, 30, 31, 33, 34, 35, 36, 37, 39, 40, 41, 42, 43, 44, 59, 60
bandgap, 59
bandwidth, 40, 42, 43, 44
beams, 3, 4, 18, 26, 33, 36, 37
binding, 35, 59

C

cell, 3, 38
character, 31, 52, 53
charge coupled device, 4
color, 40, 41
competition, 53
concentration, 49
configuration, 1
crystals, ix, 1, 4, 7, 35, 59, 60

D

degenerate, 45
dehydration, 29, 49
deviation, 21, 23
dielectric constant, 36, 43, 47
disorder, 47, 48
dispersion, 10, 31, 59, 60
displacement, 29
distribution, 4, 18, 26, 27, 29, 31, 34, 35, 36, 45, 48
doping, 4, 26

E

electron, 4, 25
emission, 47, 50, 52, 53, 54, 57
energy, 7, 8, 49, 51, 52, 53, 54
energy density, 53
experimental condition, 18, 29, 33
exposure, 3, 7, 18, 26, 28, 29, 36, 49

F

feedback, 47, 50, 53, 55
flatness, 4, 26
fluctuations, 47

H

hologram, 4, 26, 27, 28

I

image, 4, 6, 44, 49
images, 6, 35, 48
interaction, 53, 55
interference, 3, 7, 35, 36, 45, 48

L

lasers, 47, 55
lattices, 6, 7, 10, 11, 15, 28
light beam, 8, 10, 17, 18, 22, 34
line, 3, 18, 26, 30, 38, 40, 41, 50, 51, 52, 54
liquid crystals, 59
lithography, 1, 25
localization, 47, 57

M

matrix, 19, 35
Maxwell equations, 37
microscope, 4, 6
model, 3, 18, 26
molecules, 49

O

optical density, 7
order, ix, 1, 2, 36, 37, 45, 49
orientation, 4, 7, 8, 50, 51
overlap, 7, 30

P

parallel, 8, 10, 12, 15, 33
parameter, 26, 40
parameters, 39, 40, 44
permittivity, 29, 36
PET, 18, 31, 55
photoluminescence, 49
photons, 53, 57
photopolymerization, 59
physical mechanisms, 23
physics, 57
polarization, 3, 36, 42, 60
polymer, 48
polymer films, 48
production, 2
properties, 25, 35, 56
pulse, 49, 53

R

radiation, 8, 54
radius, 18, 21, 34
random media, 47
range, 1, 2, 7, 8, 11, 15, 17, 18, 19, 20, 26, 28, 31, 49
reason, 20, 42, 53
recommendations, iv
reflection, 7, 18, 20, 22, 26, 27, 28
refractive index, ix, 1, 2, 3, 4, 8, 15, 17, 18, 20, 23, 26, 27, 29, 34, 35, 36, 39, 40, 42, 43, 44, 45, 47, 49, 55
refractive indices, 17, 23
region, 6, 7, 10
relationship, 38, 40, 41, 44
respect, 3, 17, 50, 51

S

saturation, 52, 54
scattering, 35, 47, 48, 49, 50, 53, 55, 59
semiconductor, 25, 48
shape, 7, 28, 29, 39, 52
space, 11, 43, 47
spectrum, 8, 50, 51
speed, 39
stability, 55

swelling, 26
symmetry, 2, 4, 11, 29
synthesis, 1

T

temperature, 47, 49, 54, 55, 56
threshold, 50, 51, 52, 53, 54, 55
transition, 53
translation, 3, 26, 29
transmission, 7, 8, 10, 12, 15, 18, 20, 21, 22, 28, 33, 34, 49

U

uniform, 26, 53

V

variations, 41
vector, 38, 59

W

wave propagation, 59
wide band gap, 2, 4